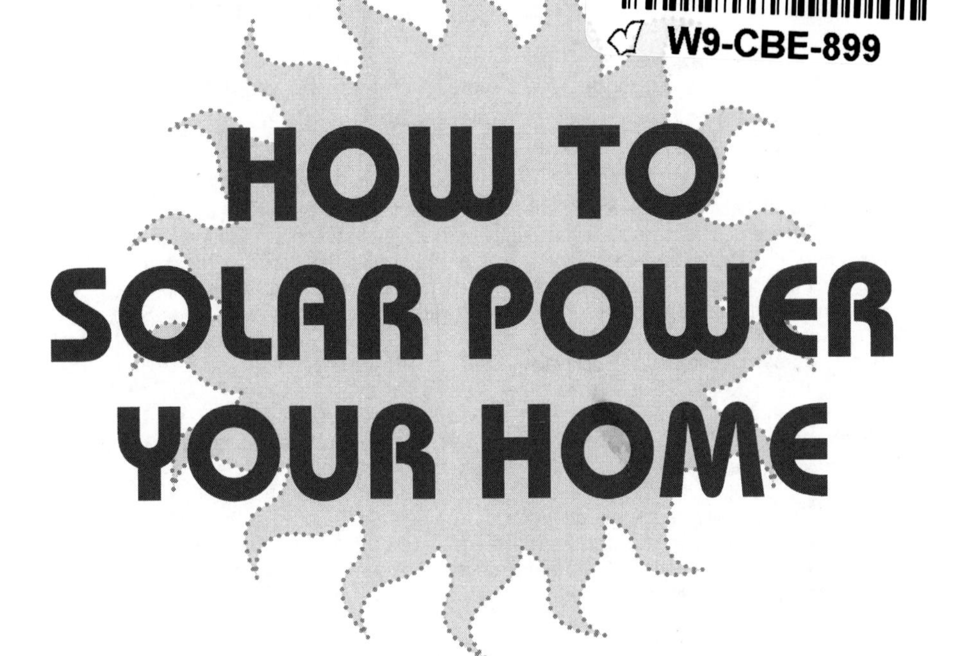

HOW TO SOLAR POWER YOUR HOME

Everything You Need to Know Explained Simply REVISED 2nd Edition

MARTHA MAEDA

How to Solar Power Your Home: Everything You Need to Know Explained Simply REVISED 2nd Edition

1405 SW 6th Ave. • Ocala, Florida 34471 • 800-814-1132 • 352-622-1875–Fax
Web site: www.atlantic-pub.com • E-mail: sales@atlantic-pub.com
SAN Number: 268-1250

Library of Congress Cataloging-in-Publication Data

Maeda, Martha, 1953-
How to solar power your home : everything you need to know explained simply / by: Martha Maeda. -- Revised 2nd edition.
pages cm
Includes bibliographical references and index.
ISBN 978-1-62023-006-0 (alk. paper) -- ISBN 1-62023-006-2 (alk. paper) 1. Solar houses--Design and construction. 2. Building-integrated photovoltaic systems. 3. Dwellings--Energy conservation. I. Title.
TH7414.M34 2015
697'.78--dc23

2015016802

EDITOR: Melissa Shortman • mfigueroa@atlantic-pub.com
BOOK PRODUCTION DESIGN: T.L. Price • design@tlpricefreelance.com
COVER DESIGN: Meg Buchner • megadesn@mchsi.com

Printed on Recycled Paper

Printed in the United States

Reduce. Reuse. RECYCLE.

A decade ago, Atlantic Publishing signed the Green Press Initiative. These guidelines promote environmentally friendly practices, such as using recycled stock and vegetable-based inks, avoiding waste, choosing energy-efficient resources, and promoting a no-pulping policy. We now use 100-percent recycled stock on all our books. The results: in one year, switching to post-consumer recycled stock saved 24 mature trees, 5,000 gallons of water, the equivalent of the total energy used for one home in a year, and the equivalent of the greenhouse gases from one car driven for a year.

Over the years, we have adopted a number of dogs from rescues and shelters. First there was Bear and after he passed, Ginger and Scout. Now, we have Kira, another rescue. They have brought immense joy and love into not just into our lives, but into the lives of all who met them.

We want you to know a portion of the profits of this book will be donated in Bear, Ginger and Scout's memory to local animal shelters, parks, conservation organizations, and other individuals and nonprofit organizations in need of assistance.

– Douglas & Sherri Brown,
President & Vice-President of Atlantic Publishing

Table of Contents

Introduction

Photovoltaic energy is electricity produced by the sun's radiation. It is considered clean, renewable energy because its production does not burn fossil fuels, release greenhouse gases into the atmosphere, or have a damaging impact on the environment. Since the 1970s, scientists, engineers, academics, and government officials have regarded photovoltaic energy (PV) as a key to solving our worldwide energy crisis and slowing climate change. Today the solar industry is rapidly expanding, with new technologies coming on the market and prices dropping every year.

The idea of using the sun's radiation to supply the energy we use every day is attractive. The concept is simple: Put some solar panels on your roof and enjoy free electricity and the satisfaction of knowing you are not harming the environment. The reality, as you will learn in this book, is more complicated. The cost of purchasing and installing solar equipment means that your "free" electricity will be more expensive than the electricity you buy from your utility company. The U.S. Department of Energy is working with solar manufacturers to bring down the cost, but in the meantime anyone purchasing a solar system must define his or her priorities in deciding how much to invest. As you read this book you will learn to evaluate how much energy a PV system will produce in your geographic area, and how much you will have to spend to get the energy you need.

Your federal, state, and local government and your utility company would all like to see you install solar panels on your home. Current tax deductions, rebates, and incentives can reduce the cost of your solar system by 30 to 50 percent, so that in just a few years you will be enjoying truly free electricity. Learn how to take advantage of the incentives offered in your area, and what you must do to qualify for these benefits. Few homeowners have the cash on hand to purchase a solar system outright; review the options for financing your solar system with grants, loans, and leases.

There are many ways to incorporate solar energy in your lifestyle with both small- and large-scale projects. Use solar power to charge your cell phone, pump water from a well, power your RV or weekend cottage, or run all your appliances 365 days a year. Learn the difference between a completely independent off-grid PV system and a grid-tie system that lets you sell electricity to your power company and buy it back when you need it. Besides producing electricity, solar energy can heat water or help keep your home warm in winter. Find out how you can build a home that uses the sun's energy in multiple ways to provide a comfortable and convenient environment for your family.

Producing solar electricity goes hand-in-hand with conserving energy. You cannot do one without doing the other. Before you begin planning your solar system, learn how to perform an energy check of your home and maximize the efficiency with which you use electricity.

Introduction

Installing a PV system requires the skills of a meteorologist, a carpenter, and an electrician. Most installations, especially those that tie into the utility grid, are done by professional solar contractors who are familiar with local building and electrical codes. Doing your own solar installation entails considerable research and study, because if you do not get the details right your whole system may not function properly. Whether you are hiring a solar contractor or doing your own installation, this book will familiarize you with every aspect of a PV system, from the pre-design stage to operation and maintenance. Learn how to select a contractor, what components you will need for your system, how to shop for solar panels and batteries, and how the installation is done. TIPS inserted throughout the book give helpful suggestions and alert you to information important to anyone installing a solar system.

At the end of the book you will find a directory of useful websites where you can learn more about each topic. You can find answers to all of your questions in the vast online library of information about solar power provided by government agencies, utilities, conservationists, solar manufacturers and dealers, solar contractors, non-profits, and individuals sharing their experiences with their own solar installations.

This book serves as an introduction and a resource for anyone interested in owning or installing a residential solar power system. By the time you finish reading it, you will be ready to embark on a solar project of your own.

CHAPTER

1

Why Go Solar?

Governments, economists, engineers, and planners agree that residential solar power systems are essential to meeting our future power needs. Solar power plants already provide electricity for communities in many parts of the world, and international bodies, federal, and local governments offer incentives to individual businesses and homeowners who install photovoltaic (PV) systems. A PV system installed in a commercial building with high energy needs will realize substantial savings relatively quickly. The short-term outlook for an individual homeowner is quite different. Investing $35,000 to replace a monthly power bill of $100 to $200 might not make economic sense to someone who only plans to live in a home for five or ten years. Even a homeowner who appreciates the advantages of solar power might not be able to afford a PV installation.

Many homeowners balk at installing solar power units because the initial cost is higher than the cost of buying electricity from a power plant month by month. If your electricity bill currently averages $150 a month, and you install a $35,000 photovoltaic (PV) unit, it will be approximately 19.4 years before you break even. The average lifespan of a PV unit is 25 years, which means that you can expect free electricity for about 5.5 years before you probably have to make a new investment in your system. Though the initial cost may be partially offset by tax credits or the opportunity to sell excess electricity to a local utility, the cost of converting your home to solar power is hard to justify in purely monetary terms. Nevertheless, an increasing number of homeowners are opting for solar power.

As concern grows about global warming and environmental degradation, increasing numbers of homeowners are opting to install solar panels on their homes, or to design new homes to use solar energy. This chapter explores the nature of solar energy and discusses several reasons why these homeowners feel strongly that solar energy is a good choice.

What Is Solar Power?

Solar power is radiation from the sun converted into electricity through the use of photovoltaics (solar panels). The sun's energy is also harnessed as solar thermal energy that can be converted to electricity or used to heat water and warm your home. The term "PV system" refers to the photovoltaic technology and equipment used to convert the sun's light into electricity.

Systems that convert the sun's energy into electricity or heat do not consume fuel as other traditional energy sources do. They also do not create pollution, produce noise, or use any moving parts. Solar power is a clean, environmentally friendly method of producing energy for your home.

The Solar Patriot, a modular home displayed on the national mall in April 2001 as part of the Solar Forum 2001, was permanently installed in northern Virginia. This "zero energy home" produces all the energy it needs using photovoltaic systems, passive solar designs, a geothermal heat pump, compact fluorescent lighting, and high efficiency appliances. It is connected to the electricity grid, but has the capability to operate at least 24 hours without electricity from the utility.

A Brief History of Solar Power

The use of solar power is not new, though it has become increasingly popular during the last few decades. According to NASA, the sun, one of the over 100 billion stars in the Milky Way, has existed for 4.6 billion years. The sun's energy comes from nuclear fusion reactions occurring deep inside its core, and NASA estimates that the sun has enough nuclear fuel to continue producing energy in the same way for another 5 billion years. Every 24 hours, enough sunlight touches the Earth to provide the energy for the entire planet for 24 years.

Ancient civilizations put the sun's energy to work in several ways. Many civilizations used the sun to tell time. Ancient Greeks and Romans used lenses called "burning glasses" to start fires by focusing the sun's rays. A burning glass is mentioned in "The Clouds," a play written by the ancient Greek Aristophanes in 424 B.C. The Greek architect Anthemius (c. 558) wrote that in 212 B.C., Archimedes, the Greek scientist and mathematician, used polished bronze shields to focus the heat and light of the sun and set fire to the wooden ships of the Romans who were besieging Syracuse. The Romans positioned their bathhouse windows to face the sun and benefit from its warmth, and the Greeks positioned their dwellings to absorb the sun's warmth. (This is considered a passive use of solar power because the sun's energy is being used without any equipment or devices.)

Photovoltaic energy

Photovoltaic energy was discovered in 1839 when Edmund Becquerel of France found that light hitting certain materials produced an electric current. In 1883, Charles Fritts developed a solar cell that only converted 1 percent of the solar energy striking it into electricity. Albert Einstein first demonstrated his interest in solar power in a paper published in 1905, and in 1921 he received the Nobel Prize for his theories on the photoelectric effect.

The first practical solar cells using the photovoltaic effect to convert energy from the sun were developed in the mid-1950s by Bell Labs. With its silicon solar cells, Bell Labs was able to achieve nearly 6 percent efficiency and its cells were demonstrated at a National Academy of Science meeting in 1954. The satellite Vanguard I, launched in 1958 in collaboration with the U.S. Signal Corps, carried a small (less than one watt) PV array to power its radios and operated for eight years. Three more satellites launched the same year, Explorer III, Vanguard II, and Sputnik 3, carried PV power systems. Though efforts to commercialize silicon solar cells faltered during the 1950s and 1960s, PV-powered systems became the accepted energy source for space applications.

In 1961, the UN held a conference on Solar Energy in the Developing World; the Meeting of the Solar Working Group (SWG) of the Interservice Group for Flight Vehicle Power was held in Philadelphia, Pennsylvania; and the first PV Specialists Conference was held in Washington, DC. Japan powered a lighthouse with a 242-W PV array in 1964.

During the 1970s, Dr. Elliott Berman, working with Exxon Mobil Corporation, designed a solar cell that brought the price of producing electricity down from $100 a watt to $20 a watt. Solar cells were used to power navigation warning lights and horns on offshore gas and oil rigs, lighthouses, and railroad crossings. They came to be regarded as a sensible alternative in remote areas that could not be connected to a power grid. In 1972, a PV system was installed by the French in a village school in Niger to run an educational TV. Several solar power companies were founded in 1975.

In 1973, the University of Delaware built one of the world's first PV powered residences, named "Solar One." The system was a PV/thermal hybrid, with roof-integrated arrays that acted as flat-plate thermal collectors. Fans blew the warm air above the array into phase-change heat-storage bins. During the day the arrays fed excess power to a utility through a special meter, and at night the system purchased electricity from the utility. ("Solar One" is also the name of the first solar-powered airplane; a solar-thermal power plant built in the Mojave Desert by the U.S. Department of Energy and the California Energy Commission; a concentrated solar power plant in Boulder City, Nevada; and a solar-powered building in New York City.)

The first PV systems providing power for entire villages were established in Schuchuli, Arizona, and Tangaye, Upper Volta, in 1979. By 1982, worldwide PV production exceeded 9.3 megawatts. Solar power units proved more cost-effective for supplying power to remote locations and villages than the installation of electrical cables to bring electricity from faraway electric power plants.

Solar power today

In 1999, according to the U.S. Department of Energy, the amount of PV installed worldwide reached 1,000 megawatts, and that amount reached 139 gigawatts by the end of 2013. Solar panels are still made of silicon but are much more efficient than in the past, with some achieving more than 44 percent efficiency. Power plants that generate electricity using solar energy are being built all over the United States. In April 2010, Florida Power & Light (FPL) opened the Space Coast Next Generation Solar Energy in Cape Canaveral to provide electricity to Florida homes. The 10-megawatt solar plant, featuring approximately 35,000 highly efficient solar photovoltaic panels on 60 acres at NASA's Kennedy Space Center, will generate energy for more than 1,000 homes and reduce annual carbon dioxide emissions by more than 227,000 tons.

Improvements to solar modules and advancements in the industry are made every year. New technologies for converting and storing solar energy have resulted in ever-increasing efficiency.

In June, 1997, the U.S. Department of Energy (DOE) announced the Million Solar Roofs Initiative (MSR), with a goal of having one million solar roofs in place in the United States by 2010. The DOE partnered with hundreds of state and local governments, industries, universities, and community organizations to pioneer the use of photovoltaic energy by individual homes and businesses. By 2006, the equivalent of more than 377,000 solar water heating, photovoltaic (PV) systems, and solar pool heating systems had been installed in the United States. MSR was replaced in 2006 by the Solar America Initiative (SAI), which aims to achieve cost parity with conventional electricity generation by 2015. These initiatives helped to increase the acceptance of solar technology, stimulate research and development, and expand the market for PV applications. In 2001, Home Depot began selling residential solar power systems in some of its home improvement stores.

Today, hundreds of companies produce, market, and install PV systems. Whether you decide to install a system yourself or hire a contractor, you will have a wide range of products and applications from which to choose.

Over 600,000 U.S. homes are equipped with solar panels

According to the ASES, in 2014, residential-grid-connected photovoltaic, or PV, installations in the United States grew to over 7,000 megawatts (MWs). The typical size for a residential installation is about 2.5 kilowatts. Divide that into 7,000 MW and the result is 2,800,000 homes now equipped with solar arrays.

Motivation for Going Solar

Ensuring an adequate supply of electricity is a primary concern of national governments and economic bodies. Without a reliable power supply, businesses and industries cannot function efficiently, and individual citizens cannot heat or cool their homes, use modern appliances, or interact with the world through television, the Internet, and cell phone devices. An ample supply of electricity is required to keep an economy thriving in today's world. It is already clear that a global power crisis threatens economic stability in both highly developed and developing nations. During the first half of 2001, Americans saw the effects of a power shortage firsthand when California experienced six days of rolling blackouts.

Developing economies are particularly vulnerable. Pakistan faces a catastrophic energy crisis that is suffocating its industry and making life unbearable for its citizens. In January 2010, desperate Pakistanis blocked roads around Peshawar in protest after they had to endure power outages of 16 hours a day during freezing winter weather. In June 2010, Iraqi security forces opened fire on a crowd that was demonstrating against repeated power cuts that have reduced their electricity supply to less than two hours a day. Nearly every state in India is facing power deficits. During the summer of 2010, angry residents attacked power utilities in Delhi, Haryana, and parts of Uttar Pradesh.

In addition to the problems caused by energy shortages, electric power plants produce large quantities of the greenhouse gas emissions that are accelerating the effects of climate change. Governments committed to reducing greenhouse gas emissions are actively seeking cleaner ways to produce energy by using solar power and wind power.

Recognizing the need to supplement existing power supplies with alternative, cleaner, and more cost-effective sources of energy, the U.S. Department of Energy began promoting the exploitation of solar power during the 1970s, with both solar power generating plants and PV systems installed on individual buildings and residences. Electricity generated at a solar power plant must still be transmitted to consumers through costly cables and transformers that can deteriorate or be damaged by a storm or natural disaster. The electricity produced by an individual PV system can be used right on the spot, without the need for any additional equipment.

Although governments and international bodies are struggling to deal with the long-term consequences of a global energy crisis, individual homeowners are deciding that solar power is a good investment.

Environmental concerns

During the 1950s, awareness began to spread that human industrial activities were doing irreversible harm to the environment and threatening our health and well-being. Today there is a widespread understanding that we are responsible for managing and conserving our resources, not only through public policy but through our individual behavior. Many people choose solar power because they know they are doing something positive to conserve energy, reduce stress on the environment, and create a cleaner, quieter home for themselves.

Solar power produces no pollution

Solar power is clean energy — it does not produce air or water pollution, nor does it produce solid waste. There are no moving parts to create noise pollution. No toxic emissions are released into the atmosphere when solar power is generated, unlike fossil fuels, which release toxic gases into the air and create solid waste. Coal, for example, produces solid waste such as ash when it is burned to make electricity. Some of this ash is recycled into concrete, but a portion of it ends up in our landfills. The American Coal Ash Association (ACAA), a trade organization, estimates 43 percent of the ash is recycled — and the United States produced about 115 million tons of this ash in 2013. That means million tons of ash went into landfills. When coal burns, gases such as carbon dioxide and sulfur dioxide are released into the air. These along with other gases emitted into the air, are causing negative consequences for our environment called global warming. Burning natural gas also releases gases such as carbon dioxide and nitrogen oxide.

An individual who chooses solar power reduces his or her carbon footprint — a measure of the impact each person makes on the environment. Approximately 81 percent of all U.S. greenhouse gases are carbon dioxide emissions from energy-related sources. According to a Pace University School of Law energy project, over a period of 30 years, an average 2-kilowatt PV system in New York State will reduce carbon dioxide emissions by 85,576 pounds, which is equivalent to the emissions released by a car driven 6,400 miles. The National Renewable Energy Laboratory (NREL) reports that a residential PV system that generates enough electricity to meet half of the home's electricity needs can avoid releasing 100 tons of carbon dioxide into the earth's atmosphere over the system's lifetime.

NOTE: The manufacture of PV cells produces environmental toxins.

Though the process of converting the sun's energy into electricity produces no waste, toxic materials and chemicals are used in the production of PV cells. Arsine, phosphine, and hydrogen are used in manufacturing solar panels. The Brookhaven National Laboratory reports that these gases are either toxic, flammable, or both, and that "because of [the gases'] extreme toxicity, even small releases may adversely affect a worker's health." These gases can be used safely when risk management and hazard control procedures are established and observed.

Solar power is renewable energy

Coal, natural gas, and oil are **nonrenewable energy** sources that cannot be replaced when they are used up. Coal and natural gas are burned in power plants to generate the electricity that a utility company delivers to homeowners via power lines, while oil is mainly used for heating homes and transportation. Solar energy, on the other hand, is **renewable energy** — it is constantly replenished as long as the sun is shining. The use of solar energy reduces dependence on coal, oil, and gas. Though there is no way to accurately predict when the world's stores of coal, oil, and natural gas will be depleted — some say they will be gone in 50 years while others claim it will take 200 years — it is clear that human beings must quickly seek alternative sources of energy.

The sun's energy is constantly available as long as the sun is shining. Batteries and other storage systems can be used to store energy to be used at night and on cloudy days, when the solar panels do not generate electricity because the sun is not shining.

Solar power reduces dependence on fossil fuels

Many people are concerned about human dependence on fossil fuels, not only because they are nonrenewable, but because the extraction, transportation, and consumption of these fuels creates environmental hazards and requires the use of energy. Competition for access to these resources causes wars and political strife. The extraction of oil and natural gas from the ground often results in environmental devastation for people living in the immediate area and releases toxins into the ground and water table. Oil refineries also produce solid waste, such as wastewater sludge, which can contain high levels of toxic materials. According to the EPA, the construction of large oil-fired power plants has a negative impact on surrounding animal and plant habitats. Coal mining is hazardous and every year coal miners are injured or lose their lives in mining accidents. There are ethical questions surrounding the equitable distribution of fossil fuels, and the excessive use in the United States and other developed economies of resources taken from countries where human rights and the well-being of less-privileged classes are not adequately protected.

On April 20, 2010, the Deepwater Horizon oil-drilling rig in the Gulf of Mexico off the coast of Louisiana was dislodged by an explosion and sank. Before the well was finally capped on July 15, 2010, 4.9 million barrels of crude oil gushed into the waters of the Gulf of Mexico, causing extensive damage to marine and wildlife habitat, covering beaches with oil and literally halting the fishing and tourism industries. In January 2010, a barge hit a tanker that was carrying oil in Port Arthur, Texas, spilling approximately 11,000 barrels of oil into the water — the worst oil spill in Texas in more than 20 years. Residents within a 28 block-radius of the spill had to be evacuated and wildlife in the area was in danger.

In 1989, after the Exxon-Valdez oil spill, the world watched on television as 11 million gallons of crude oil spread through and beyond Prince William Sound, Alaska. Horrified viewers saw dead wildlife, species threatened with extinction, and beaches blackened with oil. More than 20 years later, there is still oil in the region and in 1999, the council overseeing restoration efforts there concluded that the oil is almost as toxic as it was the first few weeks after the spill. The council also reported that the remaining oil will take decades and possibly centuries to disappear completely.

Fossil fuels themselves can be dangerous. One recent explosion caused by a natural gas leak in Connecticut killed five people and injured 12. Windows were blown out, black smoke filled the air, and Reuters reported that the force of the explosion was felt 30 miles away. Incidents such as these make people anxious to reduce their consumption of oil and natural gas.

The United States depends on fossil fuels for nearly ⅔ of its electricity and almost all of its fuel for transportation. According to the U.S. Department of Energy (DOE), "It is likely that the nation's reliance on fossil fuels to power an expanding economy will actually increase over the next two decades even with aggressive development and deployment of new renewable and nuclear technologies."

As China and India develop, they are rapidly increasing the global demand for fossil fuels. In 2010, China overtook Japan to become the world's second-largest economy. According to a report by Goldman Sachs Group, Inc., by 2025 India will be the third largest economy in the world after the United States and China. As of July 2014, China's population had surpassed 1.39 billion and India's population was more than 1.22 billion. As more and more consumers in these countries purchase cars and appliances, their use of nonrenewable energy sources will increase and the price of fossil fuels will go up. Homes and buildings powered by PV systems will not be affected by increases in the cost of fossil fuels.

an example and educating others

choose to build homes or buildings powered with PV to demonstrate their commitment to a sustainable lifestyle. A sustainable lifestyle is one that does not deplete natural resources or harm the environment. These people want their homes to be prototypes for the adoption of alternative forms of energy. Architects and engineers sometimes build solar-powered homes to experiment with new techniques for collecting and storing energy.

Economic reasons for choosing solar power

Though the $33,000 price tag for a full PV installation may seem prohibitive, solar power makes economic sense in many situations. A PV system that supplies all your power needs may not be economically feasible, but there are many smaller, less expensive applications such as solar water heaters, pool heaters, PV landscape lighting, and security lights. Solar power is often used to supply electricity for isolated vacation homes, storage sheds, lights, and water pumps that cannot easily be connected to the power grid.

Sunlight is free

Sunlight is essentially free energy, although it is not equally available to everyone. If you live in a climate that has clear sunny skies for a portion of the year, you will be able to supply at least part of your electricity needs with solar power. Some locations are not well suited for a PV system, either because of climate or because the sunlight at the proposed site is blocked by trees, buildings, or geographical features such as mountains.

Eliminates or reduces electricity bills

If your electrical bill averages $200 a month and you install a system that provides 50 percent of your electricity, you will realize a savings of $100 a month from the point that the PV system is installed. The savings from your electricity bills will eventually earn back the money you initially invested in the system. After you have paid off your system, you will continue to pay $100 per month less for electricity. Though PV panels are said to have an active life of about 25 years, many older units are still functioning efficiently after 30 years.

Calculating Electricity Bill Savings for a Net-Metered PV System

Excerpt from "Get Your Power from the Sun," a publication of the DOE – EERE (**www.nrel.gov/docs/fy04osti/35297.pdf**)

- Determine the system's size in kilowatts (kW). A reasonable range is from 1 to 5 kW. This value is the "kW of PV" input for the equations below.
- Based on your geographic location, select the energy production factor from the map below for the "kWh/kW-year" input for the equations.

Energy from the PV system = (kW of PV) × (kWh/kW-year) = kWh/year

Divide this number by 12 if you want to determine your monthly energy reduction.

Energy bills savings = (kWh/year) × (Residential Rate)/100 = $/year saved

(Residential Rate in this above equation should be in dollars per kWh; for example, a rate of 10 cents per kWh is input as $0.10/kWh.)

For example, a 2-kW system in Denver, CO, at a residential energy rate of $0.07/kWh will save about $266 per year: 1,900 kWh/kW-year × $0.07/kWh × 2 kW = $266/year.

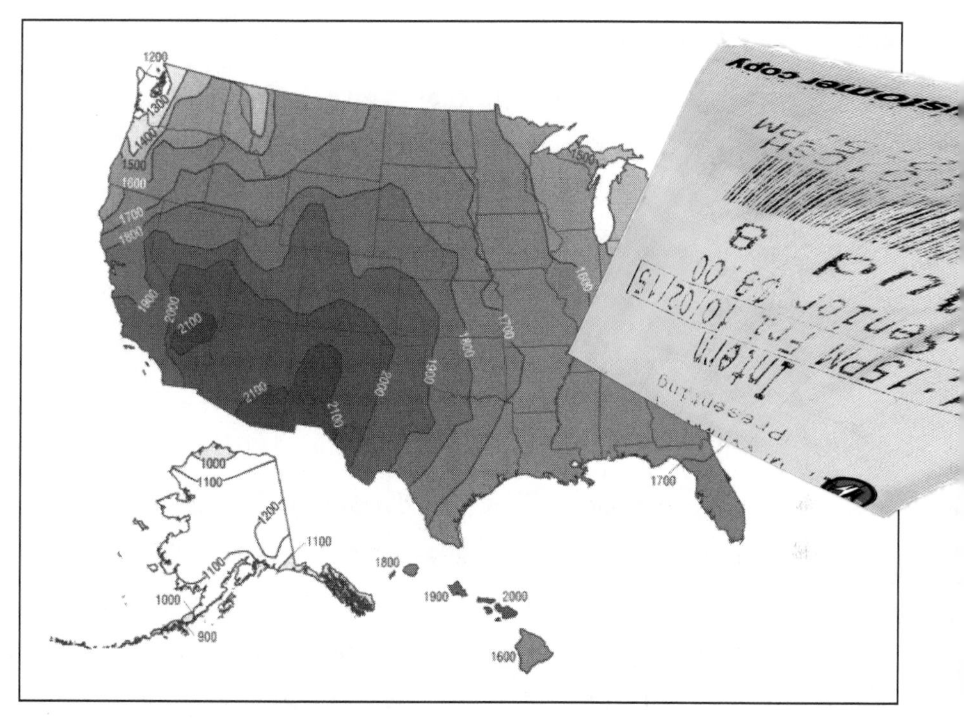

Note: The uncertainty of the contoured values is generally ±10%.
In mountainous and other areas of complex terrain, the uncertainty may be higher.

Protection from rising energy costs

Purchasing a PV system now will save you money in the future. The cost of electricity supplied by a utility is sure to increase over time because of inflation, the rising cost of fossil fuels, and the need to repair or rebuild aging power plants. When oil reached more than $136.66 a barrel in June 2008, due to a combination of supply and demand issues and market speculation, most people saw energy prices surge. It is very unlikely that your monthly electricity bill will ever go down, except when you take steps to make your home more energy-efficient. Installing a PV system now insulates you from future increases in the cost of electricity supplied by a utility.

Tax credits and incentives

To overcome financial obstacles to the use of solar power, federal, state, and local governments offer incentives such as subsidies and tax credits to homeowners to offset the cost of installing solar panel arrays that provide heat, hot water, or electricity for their homes. According to the DOE, once these incentives are applied, the cost of installing a PV system can be recouped in just five years.

Solar power increases resale value of a home

According to a report published in *The Appraisal Journal*, energy-efficient homes have an increased value linked to savings on utility bills. For every $1 reduction in annual utility bills, the home's resale value increases by $10 to $25. For example, if you save $100 a month on your electric bill, your home's value could increase by an estimated $18,000 each year, based on the assumption that its value increases $15 for every $1 of annual energy savings. If your home's value increases by $25 per $1 of utility bill savings, your home would appreciate by $30,000 when you install a PV system.

You may be able to recoup the entire cost of installing a PV system when you sell your home. Research by the American Solar Energy Society demonstrates that solar electric systems in California, in particular, increase a home's value and that the increase is often equal to, or more than, the PV system's initial cost. As utility rates rise, the PV system will cause the home's resale value to appreciate even more. Installing a PV system adds more to the resale value of a home than upgrading the kitchen or adding a swimming pool. EPA/DOE studies indicate that the value of a home increases from $11 to $25 for every $1 reduction in annual utility bills.

Practical for remote locations

Though installing a PV system may be costly, it is still an economical solution for remote locations that are far away from a power grid. Solar power plants are already used to supply electricity to some remote villages in India and Africa, avoiding the necessity of laying hundreds of miles of electric cable. PV systems are popular for isolated vacation homes and agricultural facilities, and for powering water pumps and beacon lights. You have probably observed a very common PV application — the solar panels on the telephone call boxes placed at regular intervals along U.S. highways.

When you are constructing a new home in an area that is not yet connected to the grid, the cost of a complete PV system might be less than the cost of having the power company lay cables to your house. The homeowner is usually responsible for these costs, which can range from $10,000 to $50,000 — in addition to paying monthly electric bills.

Self-sufficiency and reliability

If you have ever lived through the chaos and inconvenience of a power outage caused by a hurricane or an ice storm, you know how fragile a power grid can be. The United States experienced a particularly severe hurricane season in 2005. In the aftermath of Hurricane Katrina, the staff in New Orleans hospitals kept some patients alive by manually operating respirators after their generators ran out of fuel. After Hurricane Wilma knocked out electricity in southern Florida in 2005, the only operational gasoline pumps were at five service stations along the Florida Turnpike. Trucks could not deliver drinking water and relief supplies because they ran out of gas, and hundreds of cars lined up for miles to buy fuel for gasoline-powered generators. Cell phone service failed after 12 hours, when the batteries in Miami's cell phone towers ran out. In January 2009, a severe ice storm knocked out power to an estimated 1.3 million homes and businesses in Missouri, Kentucky, Ohio, Arkansas, and West Virginia. Hundreds of thousands of people waited for as long as two weeks while work crews struggled in the snow to repair downed power lines. Apart from discomfort and inconvenience, many people suffered economic losses because their business activities were disrupted.

Once you install a PV system on your home, you are no longer dependent on a network of aging cables and transformers to keep your computer, refrigerator, water heater, and lights on. Even if you receive back-up power from an on-grid system, a PV system will supply electricity for at least part of the day. A PV system requires some maintenance but is not as vulnerable as a power grid to severe weather and natural disasters.

Drawbacks to Solar Power

Solar power is clean, renewable energy, but it has some drawbacks that must be considered when you are deciding whether to install a PV system.

Cost of installing a PV system

This chapter has already discussed the cost of a PV installation as a major obstacle for an individual homeowner who must finance the purchasing and installation of all the system components. Even with tax deductions and subsidies this represents a considerable investment, and it will be years before the homeowner recoups the money that was spent and begins to enjoy real savings.

The cost of manufacturing solar panels largely determines what the solar panel manufacturer charges the consumer. Manufacturers are striving to lower their manufacturing costs and make solar power more competitive with other sources of electric power. In 2009, First Solar, Inc., the world's largest thin film solar module manufacturer, reported that its manufacturing capacity had grown 2,500 percent from 2004 through the end of February 2009, and that it had succeeded in lowering manufacturing costs from $1.59 per watt in 2005 to about 84 cents per watt in the fourth quarter of 2009. The goal of the solar power industry is to achieve parity — in other words, to make the cost of producing one watt of energy using solar power equal to the cost of producing one watt using fossil fuels.

Amount of energy depends on sunlight and climate

A PV system requires sunlight to produce electricity. Solar power is cheapest and most efficient in regions that experience many hours of exposure to sunlight through most of the year. In areas with a lot of cloudy weather, a PV system may not produce enough electricity to be viable. A house must have unimpeded access to a certain amount of sunlight for a PV system to be practical. If a building is situated in the shade or does not offer enough surface area to expose solar panels to sunlight, a PV system is not a good choice.

Some PV applications use lenses or mirrors to concentrate the sun's light and amplify the amount of energy that can be produced, but at the present time most of these are too expensive to be used for an individual home.

Excessive heat diminishes the capacity of a solar panel to produce electricity, so the best climate for solar power is sunny and cool. Fans and other devices can be used to dissipate excess heat, or to capture it and transform it into electrical power.

Permits and legal requirements

In most communities it is necessary to apply for permits from the local government before installing a PV system, and to comply with local zoning laws. Complex requirements may make it difficult to install an effective PV system.

If you are planning an on-grid system which will purchase back-up power from a utility and sell it excess power, you must sign a contract with the utility company. Some of these contracts do not guarantee a stable price, which means that in the future you might pay more or earn less than you expected.

The Covenants, Conditions, and Restrictions (CC&Rs) of some Homeowners' Associations place restrictions on the location and installation of solar panels on a house, because they are concerned that the appearance of solar panels may mar the aesthetic quality of the neighborhood.

Batteries and storage are needed

PV systems produce electricity only while they are being exposed to sunlight. During hours of darkness or cloudy weather, a PV system must either draw energy from a power grid, or use energy stored in some form of battery.

The manufacture of solar panels uses energy and toxic chemicals

Manufacturing PV equipment has environmental impacts that should be included in any evaluation of solar energy. Solar or PV cells use the same technologies as the production of silicon chips for computers, a manufacturing process that uses toxic chemicals. Toxic chemicals are also used in making batteries to store solar electricity through the night and on cloudy days.

The manufacture of solar panels also requires energy. A 1997 article by Richard Corkish of the Photovoltaics Special Research Centre at the University of New South Wales in Sydney, Australia, calculated that it would take somewhere between three and seven years for solar panels with thick silicon cells to produce enough photovoltaic energy to replace the energy used in their manufacture. Because most photovoltaic panels are guaranteed to function efficiently for at least 25 years, they contribute positive production of energy for 18 to 22 years. The maximum efficiency of solar crystals in 1997 was 12 percent; today some panels achieve 19 percent efficiency, reducing the amount of time it would take a solar panel to produce enough energy to compensate for the energy used in producing it.

Recycling of expired solar panels is an important component of clean energy.

Most solar panels are guaranteed to function efficiently for 20 to 25 years, and many of the earliest panels are still in use after 30 years. Eventually solar panels will need to be replaced. If solar power systems come into widespread use, as anticipated, discarded solar panels will create serious waste management problems.

First Solar (**www.firstsolar.com/en/recycle_program.php**), the largest manufacturer of thin film solar modules, has established the solar industry's first comprehensive, prefunded module collection and recycling program, designed to minimize the environmental impact of producing PV systems. As each module is sold, First Solar sets aside funds for the estimated future cost of collection and recycling in custodial trustee accounts. The site of each module installation is registered with First Solar, and labels on each module contain contact information and instructions for returning the product free of charge. When the customer is ready to discard a module, First Solar provides free packaging and transportation to a recycling center, where valuable materials are recovered and almost 90 percent of each collected First Solar PV Module is recycled into new products.

Many people who believe solar power is essential to human survival and the preservation of the environment are now choosing to adopt solar technologies. However, until solar power becomes economically competitive with other sources of electricity, the majority of homeowners will continue to balk at the cost of a PV installation. In the meantime, federal, state, and local governments are actively encouraging the adoption of solar power through financial incentives, pilot projects, research, and development of new technologies, and publicity in the media.

A complete PV system that supplies all of your energy needs is not practical in every situation. If you are interested in using solar power to supply energy to your home, you can choose from a variety of solar power technologies, or combine solar power with electricity from a utility. The following chapter describes the different ways in which solar energy can be harnessed and put to work in your home.

Active and Passive Uses of Solar Energy

Sunlight converted into heat without the use of mechanical devices, such as motors, pumps, or solar cells, is known as passive solar energy. Systems that enhance the natural collection of solar energy with pumps, ducts, or fans that circulate warm air or water are considered hybrid systems. Systems that heat water or produce electricity are referred to as active solar systems.

Though this book is primarily about active solar systems, it is important to know something about passive solar energy technologies. Passive solar energy involves designing a home or a structure to make maximum use of available sunlight and air. A successful active solar energy system also requires the efficient use of energy. Many of the passive techniques used to harness solar energy can also be used to make a solar-powered home more energy-efficient.

Passive Solar Heating

When you park your car in the sun on a summer day, the temperature inside can quickly rise to 120 or 140 degrees, even with the windows slightly open. The steering wheel becomes so hot you cannot touch it, and metal seat belt buckles can burn you. Your car is acting as a solar collector — a box with glass sides that concentrates the sun's light into heat energy. The use of passive solar heat involves designing your house as a solar collector, concentrating the energy of sunlight streaming through windows and glass panels, and storing that energy in heat-absorbent materials to be released as warmth for many hours afterwards. When you open your curtains on a cool day and let the sun shine into a room to warm it, you are using passive solar energy.

A fundamental law of physics is that heat moves from warmer materials to cooler materials until both materials are the same temperature. A passive solar energy design makes use of four basic heat-movement and heat-storage mechanisms:

Conduction — Conduction is the movement of heat through a solid material. Molecules near a heat source begin to vibrate vigorously, and the vibrations spread to neighboring molecules, transferring heat further and further from the source.

Convection — Convection is the circulation of heat through liquids and gases. As fluid or air is heated, its molecules spread apart and it becomes lighter. Warmer, lighter fluid and air rise, and cooler denser fluid and air sink. This is why heat rises to the second floor of a home while the bottom floor stays cool.

Radiation — When heat radiates, it moves through the air from warmer objects to cooler ones, like heat from a fireplace that warms the people sitting in front of it. Passive solar design incorporates two types of radiation: solar radiation and infrared radiation. When radiation strikes an object, it is absorbed, reflected, or transmitted, depending on that object's physical properties. Depending on their color, opaque objects absorb 40 to 95 percent of incoming solar radiation from the sun. Darker colors typically absorb a greater percentage than lighter colors. Bright-white materials or objects reflect 80 to 98 percent of incoming solar energy.

Infrared radiation occurs when heat radiates from warmed surfaces toward cooler surfaces. Clear window glass allows 80 to 90 percent of solar radiation to pass through into a house, absorbing or reflecting only 10 to 20 percent. After solar radiation passes through the glass and is absorbed by materials in the home, it is radiated again from the interior surfaces as infrared radiation. Although glass allows solar radiation to pass through into the house, it absorbs the infrared radiation from the interior surfaces and then radiates part of that heat back to the home's interior.

Thermal capacitance — Thermal capacitance refers to the ability of materials to store heat. Thermal mass — the materials used to build the home — changes its temperature, either by storing heat from a warm room or by converting direct solar radiation into heat. The more thermal mass, the more heat can be stored for each degree rise in temperature. A homeowner building a new home can determine how to properly position it to take maximum advantage of the sun's energy. South-facing windows allow more sunlight to enter the house. Building materials that absorb heat (thermal mass) can be placed where the sunlight will hit them during daylight hours. The interior design facilitates convection currents that move warm air to other parts of the house. At night, shutters can be drawn over windows to slow the transfer of warmth out of the house. Passive solar design also incorporates mechanisms for cooling the home during warmer weather.

Once a home has been built to use passive solar energy, very little maintenance is required. Passive solar design not only reduces dependence on fossil fuels and saves money on utility bills — it is cleaner and healthier for the occupants of the house. Warmth radiating from walls or floors is more comfortable than streams of hot, dry air blasting from vents. There is none of the dust and mold that often accumulate in the ducts of heating and cooling systems.

Owners of existing homes can make minor changes to take advantage of passive solar energy. In most parts of the United States, homeowners need to maximize solar heat in the winter and minimize it to keep their homes cool in the summer. Adding or enlarging south-facing windows will increase the amount of solar energy entering the house because south-facing windows get the most sunlight hours in the United States. Skylights can also be installed in the south side of a home to let more sun in. Thermal mass can be added by covering a sunlit wall with a heat-absorbent material such as stone or brick.

If you do not want to undertake a major home-improvement project, there are still small changes you can make to implement some principles of passive solar design. During the summer, a room facing south will get the most sun and probably will be uncomfortably warm. To diminish the amount of sun and heat entering that room during the summer, glaze windows or hang awnings outside. Outdoor shades on south-facing windows are effective in blocking the sun's heat.

Landscaping is another passive solar solution. Planting trees or hedges near your windows will help to lower your air conditioning costs. Plants and trees can also help keep your home warmer during the cold months. Take an inventory of where your home seems the hottest during the summer, usually the southern area of your home. For instance, if your family room feels like a sauna even when your thermostat is set at 76 degrees, the room may be getting too much heat from the sun to be cooled by your air conditioner. A well-placed tree outside the room's windows will help cut the amount of solar radiation entering the room and make it much cooler. In the winter, observe where the wind is strongest on your property and plant a few strategically placed trees to create a windbreak.

Solar Heating

Considering that heating accounts for 29 percent of the utility bill in a typical U.S. home, it may seem that using solar power to heat your home could result in big savings. The DOE, however, reports that in most cases it is not practical or cost effective for an active solar powering system to supply 100 percent of a home's heat. A system adequate to supply all of the heat for a home is very costly to install. At night and on days when the sun does not provide enough power, you will need a back-up power system. It would not be economical to run a back-up generator all night and on cloudy days. Most building codes and mortgage lenders require a home to have a back-up heating system, adding even more to the cost of using solar power.

Check with your local building department before you contemplate a solar heating system or a solar air conditioning unit and get information about local building codes and special permits for homes in your area. Then check the requirements of your mortgage lender. If your lender requires you to have a conventional heating or cooling system, but will allow you to have a solar power system to augment your conventional system, you can decide if you want to invest in solar.

According to the DOE, an active solar heating system that provides 40 to 80 percent of your home's heating needs is most economical. Any less than 40 percent is not cost-effective because of the high initial costs and the additional need for another heating system as a backup.

Examine your home's heating needs carefully before you commit to a solar heating system. The effectiveness of solar heating depends on many factors, including your home's heating needs and your home's geographical location. In the South or Northwest, where winters are usually milder, you require less heating than someone who lives in New England.

Two basic types of systems are used in solar energy collectors. A liquid-based system heats either water or an anti-freeze solution in a liquid-based collector. An air-based system heats air in an air collector.

The liquid-based solar heating system can supplement a forced hot air heating system, a conventional system that uses a furnace to heat air and circulates it through the home via ducts. Many homes have forced hot air heating systems. The liquid-based system is mostly used when there is storage, such as a boiler or radiator, included in a heating system.

An air-based solar heating system uses air to absorb and transfer solar energy and can heat individual rooms in your home. It can supplement forced hot air heating systems. The collectors are usually integrated into roofs or walls. Wall installations should be on an exterior, south-facing wall. The sun heats a metal plate located on the collector, which heats the air in the collector. The air is then circulated into a room via ducts from roof-mounted collectors; in wall-mounted systems, electrically powered fans or blowers are used to circulate the air.

Wall mounted solar system that heats one room

(**www.yoursolarhome.com/solarsheat.com/solarsheat/1500.html**)

An example of an air-based system that can heat one room in your home is the SolarSheat 1500G, a wall-mounted unit with a solar-powered fan that costs about $1,900. The bottom of the collector draws air from inside your room, heats it, and blows the solar-heated air back into the room through a duct in the top of the collector. The manufacturer claims it is capable of heating a room with an area of up to 1,000 square feet when the sun is shining, and it does not need an electrical hookup because it is a self-powered collector. Because the air is immediately drawn from your room, heated, and then blown back into it via the solar-powered fan, a back-up heating source may be required when the sun is not shining due to either bad weather or nighttime. This solar air collector is used to heat one room, and is not recommended for an entire home.

Solar Water Heaters and Pool Heaters

Heating water for washing dishes, taking showers and baths, and doing laundry accounts for nearly 14 percent of a typical utility bill. Using solar energy to heat water is more practical than using solar energy to heat a home. The hot water is stored in an insulated tank rather than circulating through the whole house, so less energy is required. Also, hot water is not needed 24 hours a day; water can be heated and used during the day when the sun is shining and will retain heat in a tank for several hours after the sun sets.

According to the Interstate Renewable Energy Council (IREC), Hawaii has the most solar hot water installations because energy prices are high there and the state government offers tax credits and utility rebates as incentives to homeowners. Prior to 2006, about half of all solar water heaters sold in the United States were sold in Hawaii. As of 2007, Florida and California had the next-largest numbers of solar hot water heaters. Hot water heaters are less expensive to install in Hawaii because there is little likelihood that temperatures will ever drop below freezing there.

Solar water heaters hold water in pipes contained in panels on the roof of a building. When sunlight hits the panels and the pipes, the water is heated. Two main components of solar water heating systems are storage tanks and solar collectors. Three types of solar collectors are used for residential buildings:

* **Flat-plate collectors** usually consist of copper tubes fitted to absorber plates and are typically sized to heat 40 gallons of water. This is enough for a family of two, so you need to add an additional collector for each additional two people in your family.

* **Batch collectors,** also called integrated collector storage systems, heat water in dark tubes in an insulated box. They are not recommended for cold climates because the outdoor pipes can freeze. Water is stored in the tubes until it is used; the storage tank itself is the collector. Because the hot water is held in storage until it is used, it can become extremely hot. A **tempering valve** decreases the water's temperature before it reaches a faucet.

* The **evacuated tube solar collector** is considered one of the most efficient types of collectors. These collectors work well in very cold temperatures and have long, transparent glass tubes. Inside each tube is a pipe covered with absorbent material that collects heat from the sun, containing a heat-transferring fluid. These models are efficient but much more expensive than other types of collectors.

Three types of circulation systems are used to distribute hot water from the collectors for use in the home:

- In **direct circulation systems**, water is pumped through the collectors and into the homes. These systems work best in areas where it rarely freezes, such as southern Florida.
- **Indirect circulation systems** pump a non-freezing, heat-transferring fluid through the solar collectors. The liquid absorbs solar heat, and then passes through a heat exchanger where the heat is transferred to the residence's water supply. These systems make sense in climates that are prone to freezing, such as areas in the Midwest.
- **Forced circulation systems** use electric pumps, controllers, or valves to move the water from the collector to the storage tank.

This image shows a water heater in a zero-energy house that combines active and passive solar technologies and a Freon-based geothermal heat pump. The thermal panels on a garage roof provide energy for this water heater. If more hot water is needed during a prolonged dark spell, the geothermal heat pump can heat the water.

Solar water heating systems almost always require a back-up water heating system as there will be cloudy days and days when you need more hot water than your solar water heating system can provide. A conventional water heater can serve as a backup.

A solar hot water heater can eliminate a substantial portion of your utility bill. Conventional water heaters generally have to be replaced after 15 years. If you have to replace your old water heater, which would cost you about $1,500, you could invest an additional $1,500 to install a solar water heater.

Heating your pool

Solar thermal energy can also be used to heat a swimming pool. If you do not use your pool heater very often, and if the cost of operating it is not high, you may not recoup your investment in a solar pool heater for many years. There are more cost-efficient ways to reduce the amount of energy you use to heat your pool.

Solar pool covers resembling oversized sheets of bubble wrap collect the sun's heat and transfer it to the pool water. These covers are much less expensive than installing solar panels just to heat your pool. A 24-by-40 foot rectangular cover can be purchased for about $200, while solar panels could cost approximately $5,000.

Understanding Electricity

To successfully install and maintain a PV system, you need a good understanding of what happens when you turn on a lamp, air conditioner, or hair dryer in your home. There is no simple definition for electricity. The term "electricity" refers to a number of separate phenomena related to electrical energy and the transfer of electrical energy:

Electrons — All matter consists of molecules, which are made up of atoms. Every atom consists of neutrons, protons, and electrons. Electrons have negative charges. Whenever electric current appears in a conductive material, electrons in the material break free from their atoms and begin to flow.

Electric charge — Electric charge is a basic component of ordinary matter. The atoms that make up matter are composed of neutrons, protons, and electrons. The protons and electrons are partially composed of electric charge. Electric charge cannot be destroyed; it can only be moved from one place to another, which is called "electric current." Electric charge can be stored in batteries as a build-up of electrical energy. Electric charge is measured in coulombs.

Electrical energy — Electrical energy, also called "electromagnetic energy," "EM energy," or "electromagnetic vibrations," is a type of rapidly moving wave energy. X-rays, light, microwaves, radio signals, telephone signals, the 60Hz energy generated by electric company power plants, and DC energy from batteries are all forms of electric energy with different frequencies. Energy is measured in Joules.

Electric current — Electric current exists whenever electric charge moves or flows, such as when electricity flows through a cable. Electric current is measured in amperes, or amps (A), and is represented by the symbol "I."

Voltage — Voltage causes current. It is something like a pressure that pushes electrons to start flowing when an electric circuit is connected to a source of power such as a battery, generator, or PV panel.

Electric potential — Electric potential is the potential difference in electrical energy between two points, such as the positive and negative terminals of a battery. Electric potential is measured in volts (V). The higher the voltage, the greater the capacity for work.

Electric power — Electric power refers to the "flow rate of electrical energy." Energy is measured in Joules; when energy flows it is measured in Joules per second, or Watts (W). Electric charge can be stored in batteries; power cannot. Power exists only when electrical energy is flowing.

1 Watt = 1 Joule per second

Electromagnetism — Electromagnetism is the relationship between magnetism and electricity, which enables mechanical energy in a generator to generate electrical energy, and electrical energy to generate mechanical energy, such as when electrical energy runs a motor.

Conductor — A conductor is a material or element that allows the free movement of electrons, permitting a flow of electric current. Most electrical conductors are metals, such as copper, that have free electrons in their atomic structures. Thermal conductors are materials such as metal, glass, and liquid salts that do not absorb radiant heat, allowing thermal energy to flow. Conductors are used to make the cables that carry electric current.

Semiconductor — Semiconductors are elements or compounds such as silicon or germanium that allow only a partial flow of electrons. Their chemical structure is crucial to PV systems. The atomic structures of semi-conductive materials have four electrons in their outer orbitals, allowing them to bond neatly with four neighboring atoms to form a crystal lattice. By introducing small quantities of alien substances, known as dopant atoms, with odd numbers of electrons in their outer orbitals into the material, an instability is created that allows electrons to break free and flow when they receive a small jolt of solar energy.

Resistance — Resistance is the opposite of conductivity. The resistance of a material is the degree to which it opposes (resists) the passage of electric current through it. Resistance is measured in Ohms, using an instrument called an ohmmeter. In materials with high resistance, electrons are held strongly in an energy level close to the nucleus of each atom and cannot move freely. It takes a lot of energy to displace one of those electrons.

What happens when you flip a light switch on?

When you flip on a light switch, you complete an electric circuit. Energy from a power source such as your PV system or utility initiates a flow of electrons through wires in the wall and into a light bulb, where they encounter a very thin tungsten filament twisted into a double spiral. Tungsten has a high resistance to electric current. As the agitated electrons bump against the atoms of tungsten, they jostle the tungsten electrons. The atoms of tungsten never actually let go of their electrons, but the extra jolt of energy causes the electrons to temporarily jump out to a higher energy level. As they are pulled back into their correct positions in the tungsten atoms, the displaced electrons emit energy as heat-carrying infrared light photons, a portion of which is visible light. Most of the energy is given off as heat and wasted, which is why incandescent light bulbs quickly become too hot to handle, and why governments and utilities are offering incentives to hasten the transition to fluorescent, halogen, and LED lighting that give off very little heat.

Many of your electronic devices and household appliances are operated by electric motors. Electricity enters one end of the motor, and an axle or shaft turns on the other end. Electric motors use magnets to convert electrical energy into mechanical motion. Electric current flowing through a wire generates a magnetic field around the wire. Electric motors have electric current flowing through a coil of wire wrapped around a shaft or axle in the center of the motor. Permanent magnets are placed in a casing on either side of the coil. As electric current travels through the coil of wire, it produces a magnetic field that is attracted and repelled by the positive and negative forces of the magnets, causing the coil to constantly move forward. Under normal circumstances, such a coil would reverse itself after moving half a rotation. This is prevented by the use of alternating current (AC) which regularly reverses direction, keeping the coil moving forward. Electric motors running on DC, such as toys that run on batteries, have a component called a commutator on the end of the coil that reverses the direction of the current after the coil has made a half-rotation.

How electricity is measured

Electricity is measured in terms of volts, amps, and watts.

Volts (V) measure the "pressure" under which electricity flows.

Amps (A) measure the amount of electric current. Current is the flow of electrons in a circuit and is represented by the symbol "I."

Ohms measure the amount of resistance in the material the current is flowing through. Resistance is the opposition to the flow of current and is represented by the symbol "R."

According to Ohm's Law, voltage is equal to current multiplied by resistance:

$$V = I \times R$$

Resistance can be calculated by dividing the voltage by the current:

$$R = V \div I$$

Power is a measure of the rate of energy conversion. Power is measured in watts and represented by the symbol "P." When you are shopping for an iron, a light bulb, or an electric kettle, the wattage is usually written on the package. The higher the wattage, the more power the appliance consumes.

Power = Volts × Current ($P = V \times I$)

Example: A 12-volt circuit with 8 amps of current is equivalent to 92 watts of power.

Power is also equal to the square of the current multiplied by the resistance:

$$P = I^2 \times R$$

Electrical energy is measured in watt hours (Wh) or kilowatt hours (kWh). Energy refers to the capacity for work and is represented by the symbol "E."

Energy = Power × time

Watts (W) measure the amount of work done by a certain amount of current at a certain voltage.

1 watt = 1 amp multiplied by 1 volt
1 amp = 1 watt divided by 1 volt

Watts (W) — measure the rate at which electricity is being used at a specific moment. A 100-watt bulb draws 100 watts of electricity at any given moment.

Watt-hours measure the total amount of electricity used over time. Watt-hours are a combination of the how fast the electricity is used (watts) and the length of time it is used (hours). For example, a 100-watt light bulb, which draws 100 watts at any one moment, uses 100 watt-hours of electricity in the course of one hour.

Kilowatts (kW) measure the amounts of electricity needed by a household or by a large appliance, such as a refrigerator. Electricity bills are computed in kilowatt-hours.

One kilowatt (kW) = 1,000 watts

Kilowatt-hours (kWh) — Kilowatt hours are used to measure the electricity used by a household, or by a large appliance.

One kilowatt-hour (kWh) is one hour of using electricity at a rate of 1,000 watts.

Megawatts measure the output of a power plant or the amount of electricity required by an entire city. The average size of U.S. power plants is 213 MW.

One megawatt (MW) = 1,000 kilowatts = 1,000,000 watts.

Gigawatts measure the capacity of large power plants or of multiple plants.

One gigawatt (GW) = 1,000 megawatts = 1 billion watts.

Solar Electricity

Passive solar heating and solar water heaters collect and store the sun's energy as heat. Sunlight can also produce electricity when it strikes certain types of materials. This phenomenon is called the photovoltaic effect. Photovoltaic cells (PV cells) are small units that convert sunlight into electricity. Each PV cell produces only a small amount of electricity, but groups of PV cells are connected electronically and placed in support structures or frames to form PV modules designed to produce electricity at a certain voltage, such as 12 volts. The actual voltage produced depends on the amount of sunlight striking the module. These modules are mounted in sturdy frames with positive and negative terminals, called solar panels. Solar panels are wired together to form arrays large enough to supply the voltage needed to run an appliance or supply an entire building with electricity.

How PV cells work

In 1905, Albert Einstein explained the photoelectric effect by proposing that light consists of tiny packets of energy that cause certain molecules to release electrons when they are absorbed. In 1926 these packets of energy were named "photons." PV cells are made primarily of silicon. Silicon is a semiconductor, meaning that it conducts electric current under some circumstances and not under others. When silicon absorbs

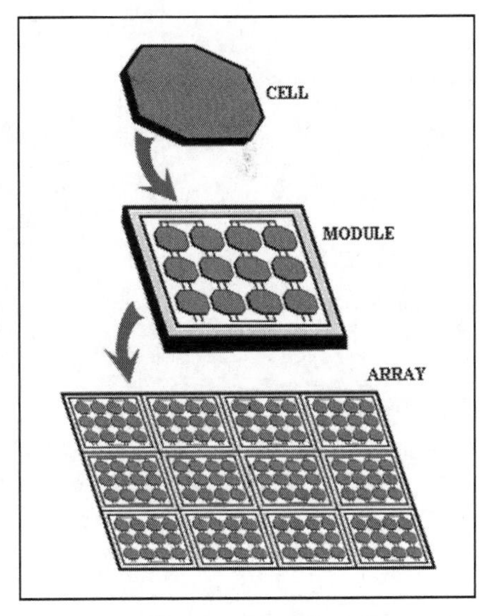

PV cells are connected together in PV modules and arranged in PV panels

photons of sunlight it releases electrons. The silicon in a solar cell is treated with phosphorous and boron to make a thin silicon wafer that has a positive charge on one side and a negative charge on the other. This creates an electric field that forces the electrons liberated by light energy to flow in one direction. Electrical conductors attached to the positive and negative sides of the silicon wafer form a circuit and draw off the free electrons in the form of electric current.

Silicon is very shiny, so it must be coated with one or more layers of anti-reflective material to prevent photons from bouncing off before they can be absorbed. A PV cell also needs protection from weather and dust, so it is typically encased in glass.

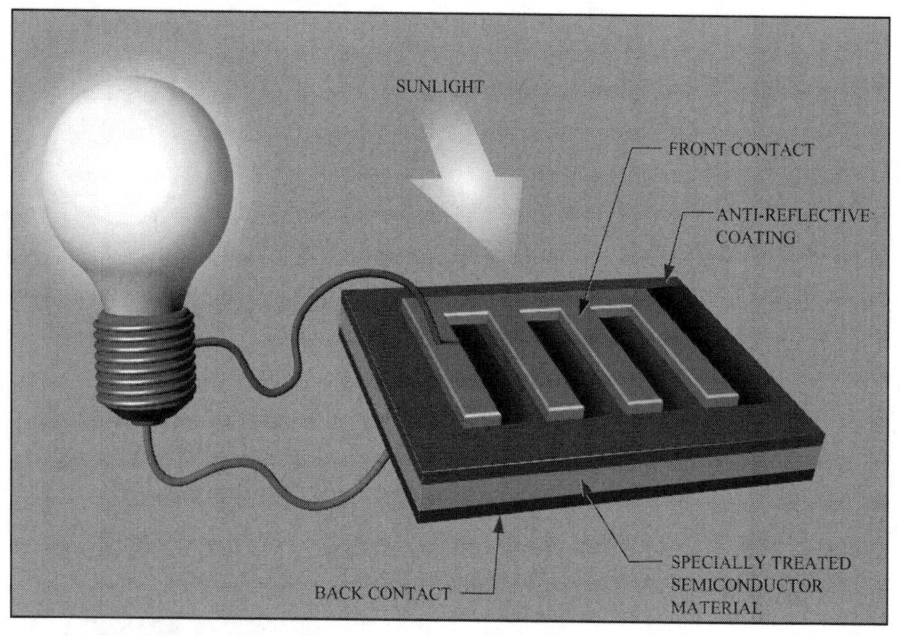

Simple diagram of a PV cell

Conversion efficiency

Much of the sunlight hitting PV cells is not converted to electricity. Conversion efficiency is a measure of the amount of sunlight that is converted by a PV cell. The conversion efficiency of the PV cells currently used in solar panels on individual residences ranges from 6 to 10 percent. PV systems used for power plants and large-scale installations employ materials and technologies that allow them to achieve a conversion efficiency ranging from 40 to 60 percent, but these are too costly for use on individual residences.

There are several reasons why a PV cell is so inefficient at converting the sunlight that strikes it into electric current. Light is composed of photons of many differing wavelengths. Some of the photons striking the surface of a solar cell are reflected and prevented from entering the cell at all. The minimum amount of energy required to release an electron from the silicon in a PV cell is known as its bandgap. Different silicon materials have different bandgaps. About 25 percent of the incoming photons have less energy than the material's bandgap and are not absorbed. When photons have more energy than the material's bandgap, the excess is re-emitted as heat or light, and accounts for another 30 percent of incoming energy that is wasted.

In some materials, a large number of the liberated electrons recombine with other molecules before being drawn into the electric circuit. Other materials contain impurities or structural defects that indirectly cause electrons to recombine with molecules. Natural resistance within the PV cell to the flow of electrons also reduces its efficiency, often at the points where electrical contacts leading to an outside circuit are attached to the cell.

Researchers are constantly striving to improve the conversion efficiency of PV cells. Different types of anti-reflective materials and surface textures are used to reduce the amount of light that is reflected by the cell. Large electrical contacts improve efficiency and resist deterioration due to exposure, but they can also block light from entering the cell. In the diagram above, a large electrical contact covers the entire back of the PV cell, but the contact on the side exposed to sunlight (front contact) is cut into grooves to expose more of the surface.

The efficiency of a solar panel is also diminished by outside circumstances. Dust on the surface of a solar panel reduces the amount of light that can enter the PV cells. This is a particular problem in desert environments where there is a shortage of water for washing the panels. Most solar panels perform best in cool sunny climates because heat reduces the efficiency of the photoelectric effect. In hot climates it is often necessary to install some kind of mechanism to cool solar panels. Finally, even a small amount of shade falling on the solar panels during part of the day greatly reduces the amount of electricity produced.

New PV technologies

The race is on to make solar cells more efficient, spurred by grants and incentives from governments and utility companies and the potential for manufacturers and energy companies to make money using new technologies. Almost every month a new breakthrough is announced. Researchers experiment with new materials, new types of semiconductors, technologies for concentrating sunlight before it enters solar panels, ways to improve efficiency by cooling, ways to use excess thermal energy to generate additional electricity, and chemical processes that use heat to release electrons. Each new technology has given rise to an entire industry as prototypes are developed, tested, and modified for commercial use. The sheer volume of information can be overwhelming to a layperson.

Most of these technologies are used for large-scale solar installations and solar power plants because they are either too expensive or too cumbersome for use on individual residences. Government agencies are partnering with private companies, however, to make new PV applications more affordable and to manufacture them in large quantities for individual use. Below is a brief overview of some of these technologies. When you begin your solar project, new types of PV cells and solar technologies may be on the market, and you should have at least a basic understanding of their similarities and differences.

One goal of solar energy research is to bring the cost of solar technology down so it can compete economically with other energy sources. Though governments would like to see millions of individual homeowners harvesting the sun's energy to run their appliances, large numbers of people will not convert to solar energy until it can clearly save them money.

Another challenge is to develop effective technologies that can be manufactured in large quantities. The silicon used to make PV cells is as common as sand on the beach, but some of the new technologies with high conversion efficiencies require rare minerals that are only available in small quantities or at great expense. Manufacturing large numbers of units employing these technologies could result in some other type of environmental degradation.

Solar cells are now being made from a variety of new materials besides silicon, including solar inks using conventional printing press technologies, solar dyes, and conductive plastics.

Thin-film solar cells

Thin-film solar cells (TFSCs), also called thin-film photovoltaic cells (TFPVs), are made by depositing one or more thin layers of photovoltaic material on a base called a substrate. The thickness of the film can vary from a few nanometers to tens of micrometers. Thin film technologies were first developed during the 1970s and are most familiar as the thin strips that power solar calculators and solar chargers for mobile devices. Until recently, thin film solar technology was not widely used on a larger scale because, though it is cheap to manufacture, its conversion efficiency was only around six percent.

New types of thin film solar cells are much more efficient, and development of thin film technology is considered a crucial component of the Solar America Initiative (SAI), the initiative launched in 2006 by the U.S. Department of Energy (DOE) to make the cost of solar energy equal the cost of conventional electricity generation by 2015. The National Renewable Energy Laboratory supports the development of thin film technology for commercial uses through its Thin Film Partnership Program (**www.nrel.gov/pv/thin_film/about.html**).

Uni-solar triple-junction amorphous-silicon solar-electric roof shingles use thin-film technology.

Thin-film technology has made it possible to create flexible solar cells and incorporate solar cells in roof tiles, glass panes, and skylights. The majority of thin-film PV units are still rigid glass panels, but the thin-film panels are much lighter.

Building-integrated photovoltaics (BIPV)

BIPVs are photovoltaic materials used in place of conventional building materials as integral parts of a building, such as the roof, facades, skylights, or windows. BIPVs became commercially available during the 1990s and have been used in many modern office buildings, factories, and homes.

The Bighorn Home Improvement Center in Silverthorne, Colorado, a high performance commercial building, features roof integrated photovoltaics, natural ventilation cooling, daylighting, diffusing skylights, and a solar wall.

Thin film solar cells are incorporated in flat roofing panels and membranes, pitched roofs, multiple roof tiles, flexible solar shingles, building facades and sidings, and windows and skylights. Transparent photovoltaics, also known as "translucent photovoltaics" because half of the sunlight that falls on them passes through, use ultraviolet radiation in addition to visible and infrared light to produce electric current. Because they replace or go over glass windows in a building, they can cover a large surface area and assist with lighting and temperature control in the building.

Awnings like this one in California provide both shade and energy.

Organic and hybrid photovoltaics

Conventional PV cells are made with inorganic silicon-based materials. Organic PV cells consist of carbon-containing molecules, known as polymers, that have a similar ability to release or attract electrons when they absorb energy from radiation. Organic solar cells are ultra-thin and flexible. They can be easily manufactured in a process that resembles printing or spraying the materials onto a roll of plastic. Organic PV cells are very inexpensive to manufacture, but they are only 4 to 5 percent efficient and must be replaced after a few years. A shortage of silicon from 2005 to 2007 raised interest in commercializing the use of organic PV cells, but the cost of producing silicon-based PV cells has since dropped. Organic PV is used mostly for specialized applications like portable chargers, solar clothing, solar umbrellas, and rollout awnings. It is also used in small building-integrated installations.

Hybrid PV cells combine positive organic materials with negative silicon-based semiconductors to improve conversion efficiency.

Multi-junction PV cells

Most PV devices in common use today have a single junction, or interface, that creates an electric field within a semiconductor such as a PV cell. Only photons whose energy is equal to or greater than the bandgap of the cell material can free an electron for an electric circuit and produce an electric current, and the remaining energy is lost. To improve conversion efficiency, two (or more) different cells, with different bandgaps and more than one junction, can be combined to generate a voltage. These are referred to as multijunction PV cells (also called cascade or tandem cells).

A multijunction device is made by stacking individual single-junction cells in descending order of bandgap (Eg). The top cell captures the high-energy photons and passes the rest of the photons on to be absorbed by lower-bandgap cells below.

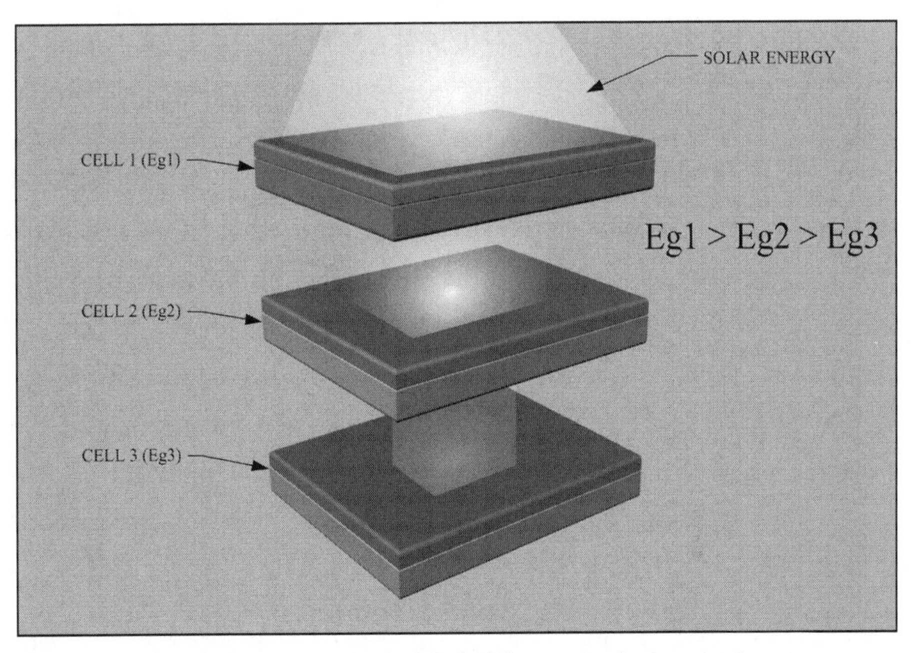

Each cell in a multijunction device absorbs photons of a different energy level, passing lower-energy photons on to the cells below.

In 2006, Boeing Spectrolabs achieved a conversion rate of 40.7 percent using a multijunction device and concentrated sunlight. In 2008, scientists at the U.S. Department of Energy's National Renewable Energy Laboratory (NREL) set a new record in solar cell efficiency of 40.8 percent using the concentrated light of 326 suns and an inverted cell made of compositions of gallium indium phosphide and gallium indium arsenide that splits the solar spectrum into three equal parts that are absorbed by each of the cell's three junctions. As of November 2009, the highest efficiency achieved under experimental conditions by Boeing Spectrolabs was 41.6 percent.

PETE (photon enhanced thermionic emission)

In 2010, Stanford scientists announced the discovery of a new PV process that converts both light and heat to electricity. Coating a piece of semiconducting material with a thin layer of the metal cesium makes the material able to use both light and heat to generate electricity. Most PV cells become less efficient when they are hot, but the PETE device will not hit peak efficiency until the temperature is well over 200 degrees C. It will perform best in solar concentrators such as parabolic dishes that can reach temperatures as high as 800 degrees. The PETE process can achieve 50 percent efficiency, and when combined with a thermal conversion mechanism could reach 55 or even 60 percent efficiency. PETE devices can be made from inexpensive materials, and a six-inch wafer is all that is required in a solar concentrator. It is anticipated that PETE devices will help solar farms to succeed in replacing other forms of energy production.

Ink-based solar cells

In 2010, researchers at IBM announced they had achieved conversion efficiency of 9.6 percent with a type of thin film solar cell made with an inexpensive ink-based process. These cells use a semiconductor material made of copper, zinc, tin, and sulfur combined with the relatively rare element selenium (CZTS). Zinc cannot be dissolved in a solvent, so particles of zinc are suspended in an ink-like slurry that is spread over a heat-treated surface. The material used in current thin-film technology includes the rare element tellurium, which could limit the quantities of cells that can be manufactured.

Nanowire solar cells

Researchers at Eindhoven University of Technology in Holland are working on solar cells made by stacking tiny nanowires (wires with diameters of 10^{-9} meters) to make a multijunction PV cell. Each nanowire in the stack can absorb a different wavelength of sunlight and release electrons to create an electric current.

A Massachusetts company, Bandgap (**www.bandgap.com**) is pioneering PV cells made of nano-silicon layers. Because the layers are so thin, Bandgap can fine-tune a PV cell to reduce reflection, increase absorption, and maximize its conversion efficiency.

Concentrated photovoltaics (CPV)

Concentrated photovoltaic (CPV) systems are primarily used in solar power plants. Solar concentrators such as lenses and mirrors focus high concentrations of sunlight on a small area of photovoltaic material to generate electricity. The concentrators are often mounted on solar trackers that follow the sun as it moves across the sky. They work best in sunny climates because on cloudy days the sun's light becomes diffused and cannot be concentrated.

Solar energy concentration is measured in "suns" — the magnitude by which the sun's radiation is multiplied. A flat solar panel receives one sun of solar radiation. For example, a concentration of three suns would be three times the radiation of the sun. CPV systems are categorized by the degree to which they concentrate the sun's radiation:

Low concentration CPV

Low concentration CPV systems have a concentration of two to 100 suns. They typically use conventional or modified PV cells, and do not become hot enough to require an active cooling system. Some low-concentration systems do not require a solar tracking mechanism because the concentrator has a wide acceptance angle — it can gather and concentrate the sunlight that naturally falls on it during the day.

Research is being done on interesting new technologies for low concentration CPV systems. Luminescent solar concentrators are plastics that concentrate the sun's radiation at an estimated concentration of two suns in a single small area, where it can be converted to electricity by a multijunction PV cell. Solar panels made of these plastics require only one PV cell and are less expensive to manufacture than solar panels made up entirely of silicon PV cells.

Researchers at the Massachusetts Institute of Technology (MIT) have developed a mixture of dyes that can be painted onto a pane of glass or plastic. The dyes absorb sunlight and then re-emit it within the glass in a different wavelength of light, which then tends to reflect off the interior surfaces of the glass. As the light reflects within the glass pane, much of it is channeled to the edges of the glass where it is emitted at an estimated 40 times its original strength. PV cells optimized for this concentration can be mounted along the edges of the glass pane to produce electricity. Covalent Solar is working to produce a commercial application of this technology.

Medium concentration CPV

Medium concentration CPV systems have concentrations of 100 to 300 suns, and require two-axes solar tracking to keep the concentrating mechanism always pointed at the sun. They also require cooling mechanisms.

High concentration photovoltaics (HCPV)

HCPV systems employ concentrating optics such as dish reflectors or Fresnel lenses to concentrate sunlight to intensities of 300 suns or more. Because of their complexity they can be used only for large-scale projects.

Four newly designed solar power collection dishes called Suncatchers were unveiled at Sandia's National Solar Thermal Test Facility.

Concentrated solar power (CSP)

Technologies to collect the sun's energy and use it to generate electricity have been in use since the 1970s. Various types of devices, known as solar concentrators, are used to concentrate the sun's radiation, store it as heat, and use it to run generators. Some of these technologies are now being used in conjunction with PV devices to concentrate the sun's radiation, draw off heat and convert it to electricity, and store energy as a backup to be used at night and on cloudy days.

Parabolic trough

Parabolic troughs, the oldest CSP technology, have been used at power plants in California and Nevada since the 1970s. Long rows of parabolic mirrors focus the sun's energy on pipes filled with a working fluid such as water, superheated steam, or molten nitrate salts. The fluid absorbs and stores the energy as heat, which can then be used to produce electricity in steam-powered generators.

Concentrating Linear Fresnel Reflector (CLFR)

Line-focus concentrating solar collectors generating power in the Mojave Desert.

CLFRs use many thin strips of mirror, arranged to concentrate the sun's energy on just two tubes of working fluid. The flat mirror strips are much cheaper to manufacture than parabolic mirrors, and can be arranged so they collect more of the sunlight. Both CLFRs and parabolic troughs can be used to increase the efficiency of existing fossil fuel power plants.

Power tower

Power towers use a technology similar to parabolic troughs, except that the mirrors are arranged all around a tower to focus the sun's light on a receiver at the top of the tower, where a working fluid is heated and used to generate electricity. When the sun's energy is captured as heat in a fluid it can be stored

LEFT: Artist's concept of the solar power plant built near Barstow, CA.
RIGHT: 300-foot tower surrounded by sun tracking mirrors at Solar Two power plant, Daggett, CA. Solar Two's 1,926 heliostats focus sunlight onto a receiver near the top of the tower, which is filled with a molten salt mixture that collects and stores enough thermal energy to drive a steam turbine that produces electricity sufficient to power 10,000 homes.

until it is needed. Two pilot plants, Solar One and Solar Two, were built and tested in California, and a third, Solar Tres, is under construction in Spain.

Solar chimney

Solar chimney technology uses several square miles of greenhouses with transparent roofs to collect solar energy. The greenhouses are built on ground that slopes up towards a large chimney. As the air heated inside the greenhouses rises up through the chimney, it turns a wind turbine. Water pipes on the floor of the greenhouses collect and store heat to operate the chimney at night and on cloudy days.

Dish/Stirling

Dish Stirling solar power system at the Arizona Public Service Solar Test and Research Center.

A dish/Stirling system consists of a parabolic mirror that focuses heat directly onto a Stirling engine. The engine is powered by a gas chamber connected to a piston and drive shaft. When the gas is heated, the drive shaft powers a turbine. Dish/Stirling systems can convert about 25 percent of the sun's energy into electricity.

Photovoltaic energy for your home

Experts agree that installing a complete stand-alone PV system to supply all of the electricity for your home does not make economic sense unless you live in an isolated location far from the power grid. It is even more costly and impractical if you do not live in a sunny climate, because you will need extensive battery storage and back-up generators. Most of the PV systems used on homes today supplement electricity supplied by a utility company. They are tied to the power grid, so they can draw power at night and at other times when the PV panels do not produce enough current. There are also many ways to use photovoltaic energy to power individual appliances or replace environmentally unfriendly acid batteries. If you are not yet ready to install PV panels to power your home, there are still a number of smaller projects using solar energy that can reduce your dependence on the power grid and cut down your utility bills. Tiny PV cells have been in use for decades in solar-powered calculators. You may already use small-scale PV technologies in solar-powered flashlights, emergency radios, and portable phone chargers.

Stand-alone, or off-grid systems

Stand-alone systems are completely independent of the power grid. When the sun is shining, PV panels generate electricity that is stored in batteries to be used when it is needed. A back-up generator fueled by fossil fuels may be needed to supply additional energy during times of the year when there is not enough sunlight. Stand-alone systems are an economical choice for homeowners living in remote rural areas who might have to pay thousands of dollars to connect to the electrical grid. They are also economical for vacation cabins and work sites.

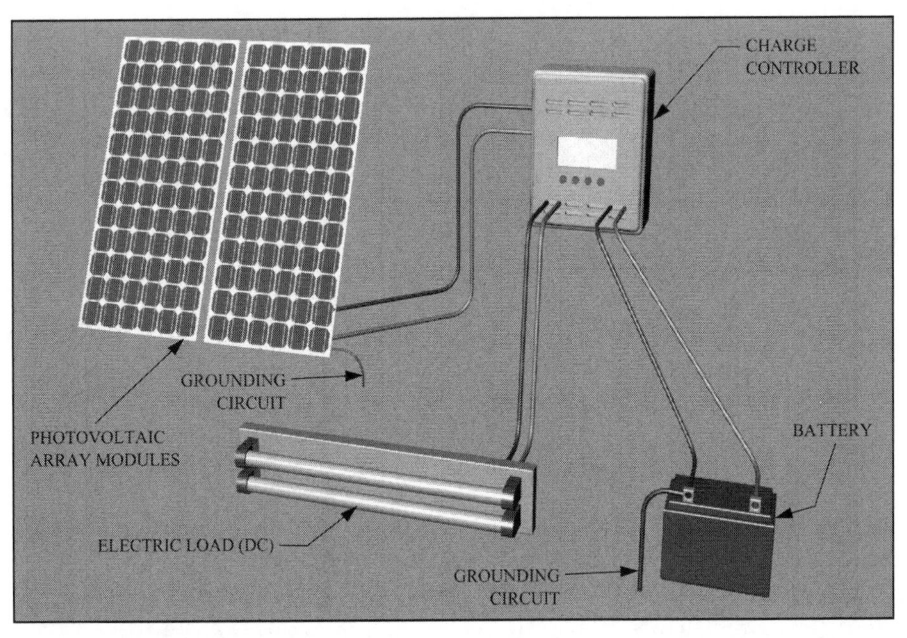

Simple off-grid system with DC only

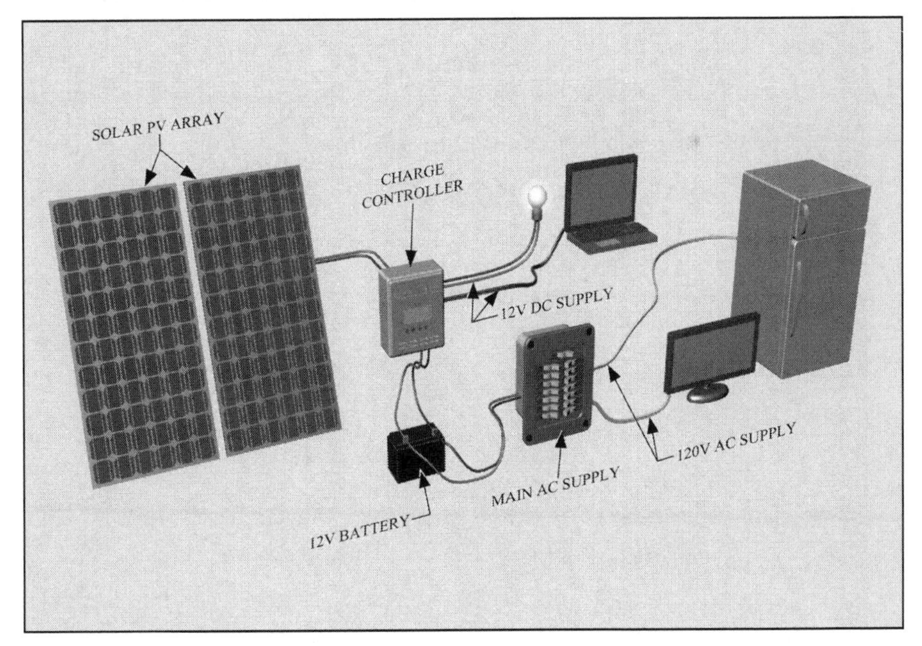

Off-grid system with both DC and AC

Photovoltaic power can be used to pump water for livestock.

Stand-alone systems can also be used for smaller projects, such as supplying power for outdoor lighting and security systems, running a pump to water livestock, or providing a back-up power supply for a business office. You have seen stand-alone PV units powering train signals and call boxes along highways.

Grid-tie, or ongrid systems

The most common systems installed on residential buildings are grid-tie PV systems. These systems allow you to use the electricity generated by your PV system while the sun is shining and buy electricity from a utility at night, on cloudy days, or at times when your energy needs exceed the capacity of your system. If your PV system generates more electricity than you need, the excess is directed back to the power grid and bought by the utility company through net metering — a program in which the utility company deducts the amount of outgoing power from the amount of power flowing in through your electricity meter.

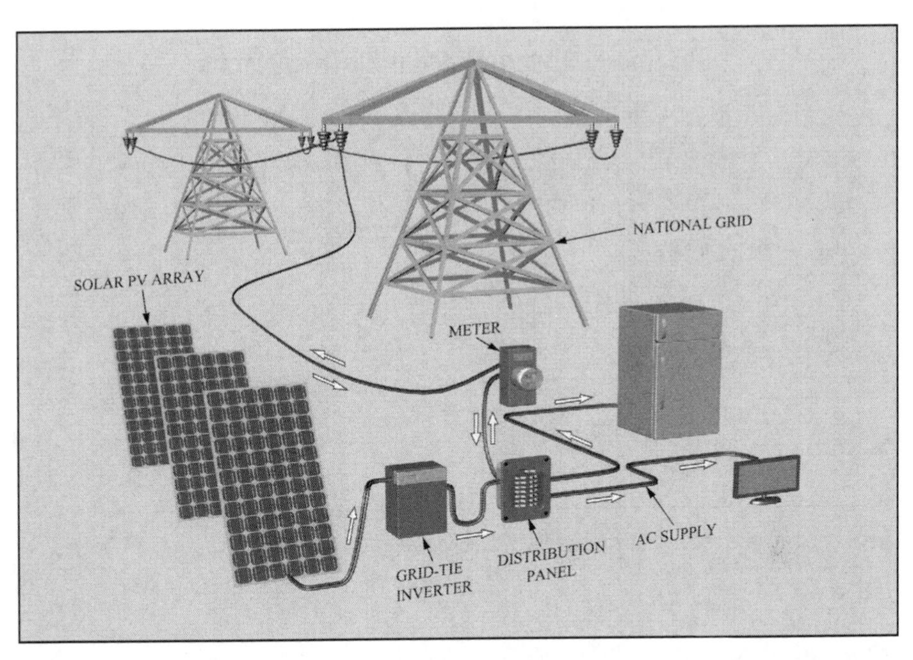

Diagram of a grid-tie system.

Many states and counties offer subsidies and tax incentives to homeowners who purchase ongrid PV systems because they relieve some of the demand on overburdened utilities.

Photovoltaic panels installed on the roof of a home.

Grid-tie works well in hot sunny climates where the use of air conditioning causes peak demand during the hours when the sun is shining.

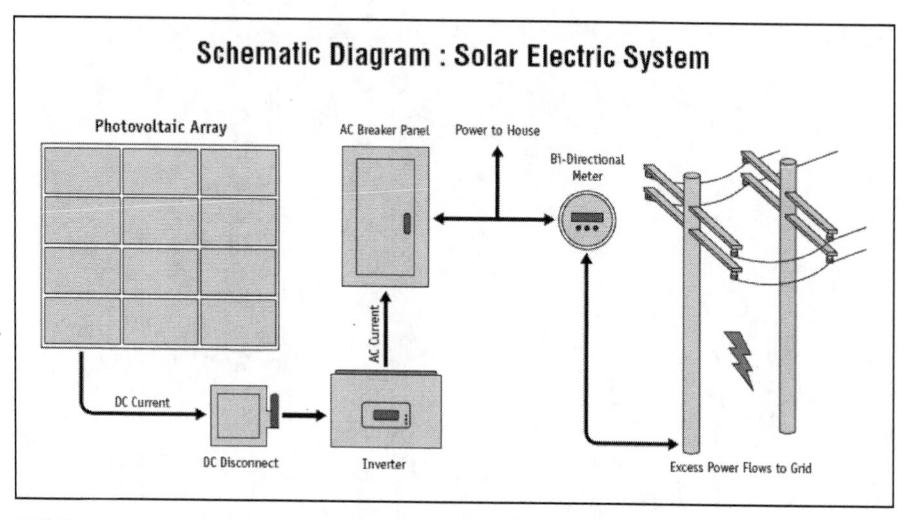

Grid-tie system sends excess power to grid and draws power when solar energy is not being produced.

Ongrid system with batteries

One drawback is that in most grid-tie systems, the power from your PV system will also be shut off when the grid experiences a power outage. A grid-tie system with a backup or grid failover system includes a bank of batteries that provide electricity during a power outage. Back-up batteries are an expensive component of a PV system, and have to be replaced every few years, so unless your area experiences frequent power outages, the extra expense may not be justified. Grid failover systems are useful in areas where the supply of electricity is unreliable, or when an uninterrupted supply of electricity is needed — for example, to refrigerate lab supplies, keep computers operating, or run medical equipment.

Grid fallback systems

In a grid fallback system, the power generated by your solar array is stored in a bank of batteries. The electricity to run your home is drawn from these batteries until they are depleted, then you switch over to the grid until your batteries are recharged. Grid fallback systems are cheaper than grid-tie systems for small household PV installations because both solar panels and batteries are low voltage units (12v, 24v, or 48v). Grid-tie systems typically require the generation of several kilowatts of electricity, which means that a larger number of solar panels must be linked together to produce a voltage of several hundred volts. A grid-fallback system that powers one or two electric circuits in your house can be built for several hundred dollars, while the least expensive grid-tie system costs thousands of dollars. If your solar array generates less than 1 kWh, a grid fallback system is probably cheaper. If your array generates more than 1 kWh, a grid-tie system may be more cost-effective.

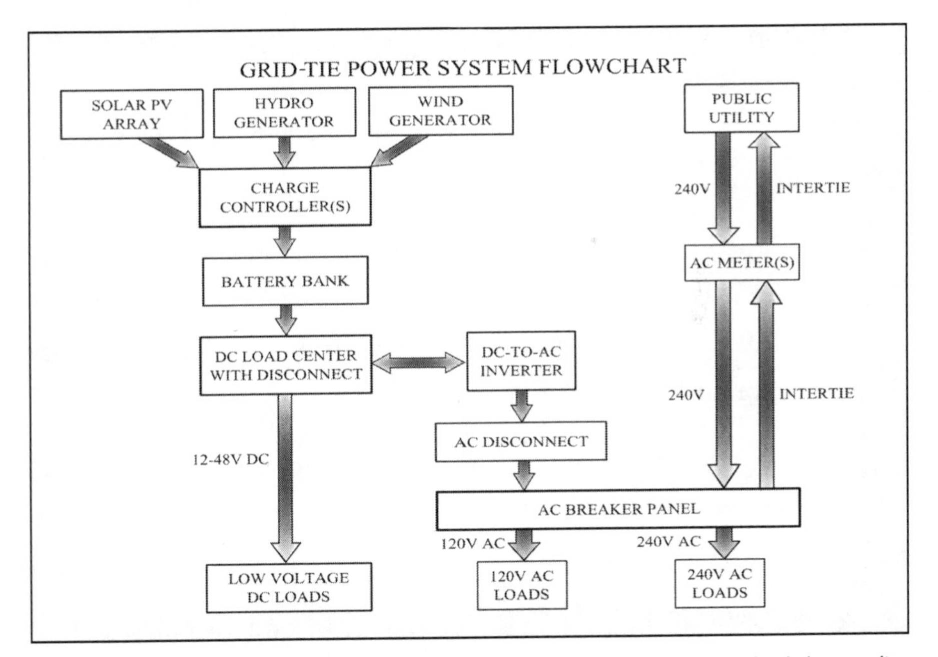

In a grid fallback system, electricity generated by the solar panels is stored in a battery bank that supplies the house. When the batteries are depleted and not being charged by the solar panels, they are charged from the grid.

A grid-fallback system can be easily expanded as more funds become available; it is difficult to expand a high-voltage system unless the expansion was part of the original design.

Grid-fallback systems do not sell excess energy to a utility — it is stored in batteries for your use. One advantage of a grid-fallback system is that you are using your own solar power even at night, as long as your batteries are not depleted.

Photovoltaic appliances and gadgets

PV technology is already incorporated in a variety of small electronics and applications that not only reduce your use of electricity generated with fossil fuels or disposable batteries, but liberate you from wall sockets and extension cords. Innovative designers are introducing new products every month. The website Ecofriend.org (**www.instadaily.com/pv/538771**) reviews many of these applications:

* Backpacks and bags with solar panels that charge cell phones, iPods, MP3/MP4 players, digital cameras, PDA, DVD players, and GPS devices anywhere you go (**www.ecofriend.org/entry/10-solar-powered-bags-to-charge-your-luggage**).
* Charging stations for electric vehicles. Some designs use solar panels to provide shade in parking lots and generate electricity for nighttime lighting.
* Solar powered speaker systems, iPod docks, and DVD players.
* Stand-alone streetlights.
* Solar watches and clocks.

* Solar radios.
* Solar powered gate openers.
* Solar powered landscaping lighting and Christmas lights.
* Solar powered security systems.
* Collapsible solar arrays and chargers that can be folded up and carried in a backpack (**www.solarhome.org/powerfilmf15-3005 wattfoldablesolarcharger.aspx**).
* Portable military field shelters that provide power for radios, communications equipment, and chargers (**www.solarhome.org/ militarysolarquadrant190watts.aspx**).

Components of a PV system

Solar panels made up of PV modules are only one part of a PV system. Depending on whether you have a grid-tie or an off-grid system, several other components are needed to store, regulate, and convert DC current from the solar panels into AC current to run the appliances in your home, and connect you to the power grid.

Solar panels

Solar panels are the starting point of a solar electric system; they are where the sun's radiation is converted into electric current. Different types of PV technology are discussed above. The solar panels used on residential buildings vary in thickness, weight, and size and may be mounted on the home's roof or exterior, incorporated in architectural features, or arranged in installations on the ground in a sunny location. They can be stationary or equipped with mechanical solar trackers that move them so they follow the sun's path throughout the day. Contacts attached to each solar panel draw off the electric current so it can be conducted through a cable.

Solar panels are typically designed to produce between 14 and 16 volts when they are under load — connected to a battery or inverter that draws off power. This means that a solar panel is capable of charging a 12-volt battery.

Solar panels can be linked together in an array to generate more power or to run the PV system at a higher voltage. An array with panels connected in series runs at a higher voltage, commonly 24V or 48V. An array with panels connected in parallel produces more power at a lower voltage. Two or more series can be wired in parallel to produce a desired voltage.

Solar array mounted on a garage roof.

Inverters

Solar panels generate low-voltage direct current (DC), but electricity supplied by the grid is a high-voltage alternating current (AC). You can purchase appliances that run on 12V DC current, but conventional electrical equipment and household appliances run on AC. If you are connecting to the grid or running AC equipment, you need an inverter, a device that converts DC to AC and steps up the voltage.

Batteries

The power generated by solar PV systems fluctuates according to the brightness of sunlight falling on the panels. Most electronic equipment needs a steady current to operate efficiently, so the system must have a way to supplement the power coming from the solar array when it drops below the required level. In a grid-tie system, the inverter supplies extra power from the grid when the PV system is not supplying enough, and draws off excess current when it is not needed.

Stand-alone and grid fallback systems use batteries to store energy and provide a constant source of power to run equipment and appliances. The lead acid deep-cycle batteries commonly used with PV systems are something like car batteries, but they are designed to be depleted and recharged over and over. Most of these are 6V or 12V batteries, and just like solar panels they can be wired in series to increase the capacity and voltage of the system, or in parallel to increase capacity while voltage remains the same.

Other types of energy storage are also used with PV systems, such as hydrogen fusion and thermal storage.

Controllers

If you are using batteries in a stand-alone or grid-fallback system you will need a controller, a device that manages the flow of electricity into and out of your batteries. Batteries will be damaged or destroyed by overcharging, or if they become completely discharged. A controller will stop the incoming flow of electricity when the batteries are fully charged, and shut off circuits before batteries are completely drained.

Brackets and mounts

Your PV panels must be able to withstand high winds, rain, snow, and intense heat for 25 or 30 years, which means they should be mounted securely, using weather-resistant hardware. In hot climates, there must be enough space underneath the panels to allow air circulation to assist with cooling. Brackets and mounts must hold the panels at the optimum angle to receive the maximum amount of sunlight. The mounts supplied by solar panel manufacturers are a good choice because they are designed specifically for those panels.

Solar tracking devices

Some PV systems incorporate mechanical solar tracking devices that move the PV panels to follow the sun's movement across the sky and receive the maximum amount of sunlight. Solar tracking devices add to the cost of installing a PV system, and at the present time are not economically viable except in situations where there is very little space to install solar panels.

Appliances and electrical devices

The final component of your PV system is the electrical devices that are going to use the electricity generated by your solar panels. Collectively known as the "load," these devices determine the amount of power you need and consequently the type of system necessary to produce it.

Conclusion

The concept of using clean energy from the sun to power our modern lifestyle is exciting. A worldwide body of scientists, researchers, and entrepreneurs are devoting their careers to developing ways to use solar energy in the hope that it can slow climate change and global warming, liberate human beings from dependence on nonrenewable fossil fuels, and provide people living in the most disadvantaged regions of the world with basic human needs.

After reading this chapter, you know that PV technology is still in its pioneering stages and has limitations that restrict its use in some situations. Your geographical location and financial resources might make it impractical for you to get completely "off the grid," but there are many ways to use PV technology to supply at least some of your energy needs. The following chapters will help you to evaluate how you use energy and take you through the steps of a successful solar energy project.

CHAPTER

3

Is Solar Energy for You?

This is the age of solar energy. According to the Solar Energy Industries Association (SEIA), U.S. homes with rooftop solar panels produced 156 megawatts of electricity in 2009, more than double the 78 megawatts produced in 2008. Growth was fueled by U.S. government tax incentives allowing homeowners to deduct 30 percent of the total cost of a solar panel installation from their taxable income, and by a drop of more than 40 percent in the price of PV modules. Solar energy still provides less than one percent of U.S. electricity. The U.S. ranks fourth, behind Germany, Italy, and Japan, for the number of solar installations. California has the most solar-electric installations of any state in the U.S., followed by New Jersey, Florida, Arizona, and Colorado.

Many national and state governments are devoting significant resources to promoting the use of solar energy to produce electricity both in power plants and in individual businesses and residences. They regard the use of solar energy as crucial to reducing greenhouse gas emissions from power plants burning fossil fuels. They also hope to defuse a looming energy crisis by decreasing individual consumers' demand for electricity delivered by utilities. Using electricity supplied by on-site solar systems also cuts down on the amount of energy that is wasted when electricity is transmitted over long distances.

How can you become a part of the solar revolution? A number of factors, including your budget, your geographical location, and your energy requirements, will determine which PV system you choose. You might decide on a full-scale off-grid PV installation, a grid-tie system that supplies part of your electricity, or a PV array that runs a specific appliance. You can start small and add more solar panels as money becomes available. You can also supplement a PV system with passive solar applications that conserve energy and maximize the natural effects of air and sunlight. This chapter will help you evaluate your individual circumstances and your energy needs, and decide what type of solar project you want to undertake.

Location and Climate

Your geographical location determines not only the amount of sunlight available to produce electricity, but also how much electricity you need at various times of the year and the conversion efficiency of your PV system. For example, if your home is in central Pennsylvania, you will receive less than 2 kWh of radiation per square meter in January, when you need electricity to heat your home. In August, Phoenix will receive between 5 and 6 kWh of radiation per square meter, during the time period when air conditioners are needed night and day. The conversion efficiency of PV panels diminishes, however, in extreme heat. The two maps below show the difference in the amount of radiation received in different parts of the U.S. in January and August. A grid-tie system will buy electricity from a utility when the solar panels cannot produce enough. An off-grid system will have to be large enough to produce the required amount of electricity, and have adequate battery storage or a back-up generator to supply electricity when there is not enough radiation from the sun.

You can estimate how much electricity a solar panel on your home can be expected to generate by looking up the solar irradiance data for your area. The amount of solar energy — number of hours of sunlight combined with the strength of the sunlight — falling on one square meter (9.9 square feet) is known as the insolation of a particular location on the globe. Insolation is expressed as average solar irradiance — ultraviolet, visible, and infrared radiation — falling on a square meter in kilowatt-hours per day (kWh/m^2/day). You must look at the solar irradiance for each month of the year. The maps below are from the Renewable Resource Data Center of the National Renewable Energy Laboratory (NREL) (**http://rredc.nrel.gov/solar/old_data/nsrdb/redbook/atlas**). They show the average monthly irradiance of the U.S. for the months of January and August.

U.S. Solar Radiation Maps

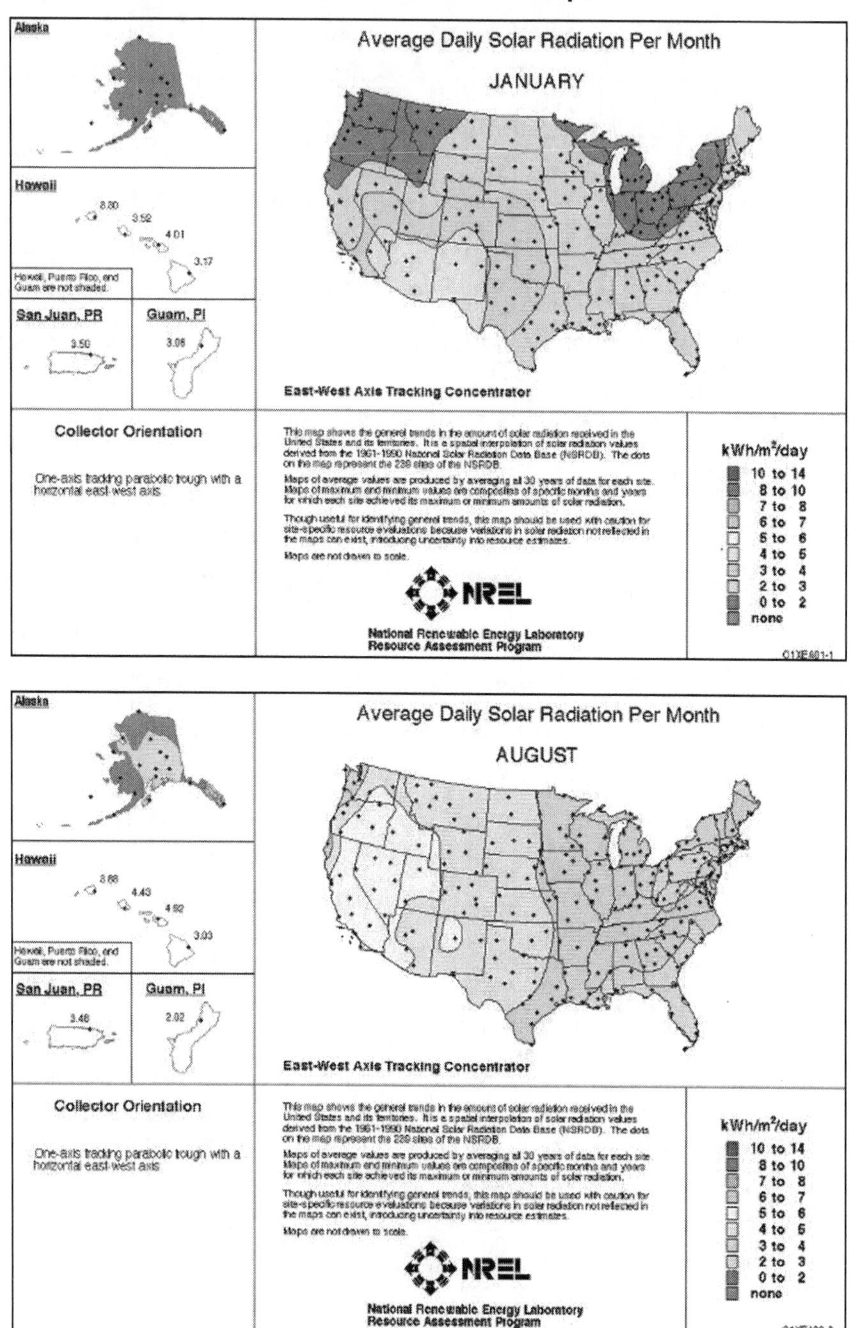

Every solar panel bears a label stating its wattage — the number of watts it can be expected to generate based on a solar irradiance of 1,000 watts (1 kilowatt) per hour. To get a rough idea of how much electricity you can expect a solar panel to produce each day, you can multiply the average daily solar irradiance times the stated wattage of the panel. For example, a 20-watt solar panel could be expected to produce 40 to 60 Wh (Watt-hours) in central California in January, and 100 to 120 Wh in August.

These calculations are only approximate, because the solar irradiance of your particular location will be affected by local weather conditions, the angle, and location of your solar panels, and whether obstacles such as tall buildings or mountains block sunlight at any time of the year.

To get an accurate idea of what to expect from your solar system, you must also look at the maximum and minimum amounts of irradiance for various times of the year. For example, during a prolonged winter blizzard or a period of cloudy days in summer, your solar panels would receive much less than the average amount of radiation. You would either need a back-up power source to meet your energy requirements, or you would have to install a system large enough to generate enough electricity even on those days.

If you live in the city with the most sunny days, Yuma, Arizona, you can expect to have 242 sunny days in a year. Phoenix, Arizona, comes in second with 211 sunny days, and Las Vegas, Nevada, comes in third with 210 sunny days. Using a PV system to supply electricity for a home would be much less practical in locations with few sunny days per year such as Cold Bay, Alaska (20 sunny days), St. Paula Island, Alaska (18), and Hilo, Hawaii (36). Regions at high latitudes experience only a few hours of sunlight on winter days and exceptionally long summer days.

You must also look at the highest and lowest temperatures recorded in your area, because your equipment must be able to withstand these extremes. If you live in a region subject to hurricanes, heavy winds, or ice storms, you need insurance adequate to cover possible damage to your solar panels.

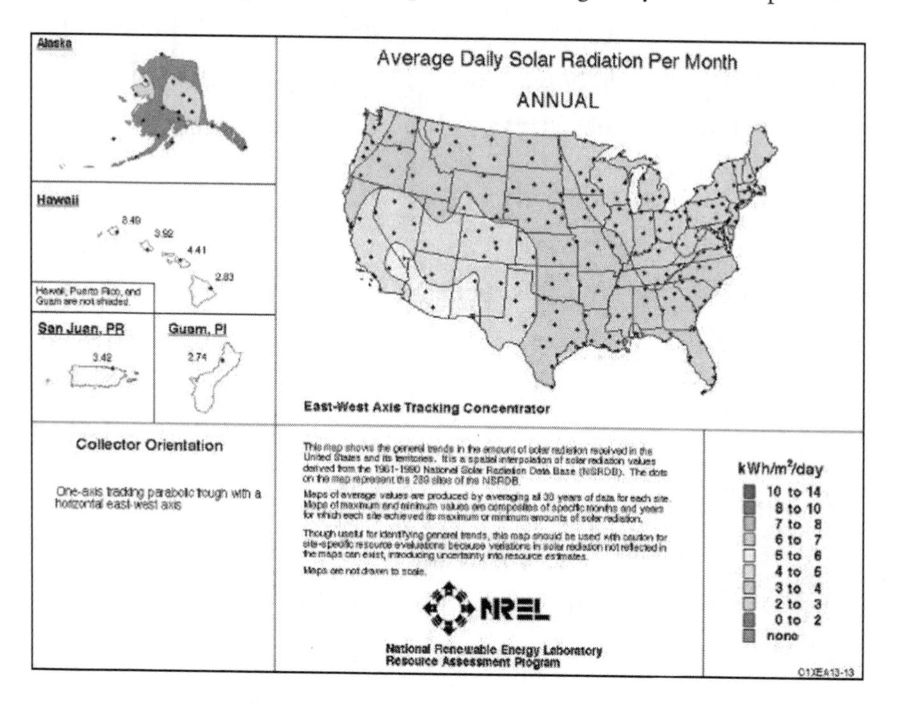

The third map above shows the annual average daily irradiation per square meter for the United States. The maps above are based on temperatures averaged over 30 years. They do not account for the fact that the earth's surface appears to be warming at a rate of about 0.29oF per decade or 2.9°F per century, according to National Oceanic and Atmospheric Administration's (NOAA) *2008 State of the Climate Report* and the National Aeronautics and Space Administration's (NASA) *2008 Surface Temperature Analysis*, a phenomena that might indicate larger amounts of radiation striking the earth. Since the mid 1970s, the earth's average surface temperature has warmed about 1°F, and the eight warmest years on record (since 1880) have all occurred since 2001, with the warmest year being 2005.

Evaluating Your Energy Consumption

U.S. homes have three direct sources of energy: electricity, fossil fuels such as coal and oil for heating, and natural gas for cooking and heating.

Chapter 2 explains that using a solar electric system to heat a home in winter is impractical because a very large solar array would be needed to generate enough power on cloudy winter days. Solar heating systems that use solar energy to heat air or liquid that transfers warmth to the interior of the home can provide 40 to 80 percent of a home's heat. Passive solar heating, wood fires, and geothermal energy can also replace fossil fuels as a source of heat in the winter.

To calculate the payoff of using a PV system to replace or supplement electricity supplied by a utility, you have to look at your consumption of electricity.

Understanding your electricity bill

Energy is measured in British thermal units (Btus). A Btu is equal to 250 calories, the amount of energy required to raise the temperature of a pound of water by 1° Fahrenheit. This is approximately the amount of heat produced by lighting a match. A Btu is such a tiny unit of energy that larger quantities of energy are measured in quads — one quadrillion Btus. According to the DOE, an average U.S household consumed 4.7 quads of electricity in 2009.

Electric power is measured in watts. Your utility company bills you for the number of kilowatt-hours (kWh) of electricity you use in a month. One kilowatt-hour (kWh) is one hour of using electricity at a rate of 1,000 watts. One kWh equals 3412.1416 Btus. When you want to compare the cost of electricity produced by a solar system to the cost of electricity from your utility, you must look at the cost per kWh.

Gather your electricity bills from the past year and add up the total number of kWh for the year. If you can view your account online on your utility company's website, you may be able to view a summary for the year. To discourage customers from using more than the average amount of electricity, most utility companies apply a baseline price for the first 1,000 kWh and a higher rate for any additional kWh. Some companies charge more for electricity during peak hours (time of use, or TOU billing), or offer a rebate if you sign up for a program that temporarily shuts off power to certain appliances when demand for electricity is high. There may also be two separate charges for "fuel" and "energy" per kilowatt-hour. Your electricity bill also contains additional charges for taxes and other fees. Ignore all of this for the time being and just divide the total amount you paid for electricity by the total number of kWh. This will tell you how much you paid per kWh for your electricity.

The table below, using data from the DOE, shows how much the average customer paid per kWh for electricity provided by utilities in June 2009 and 2010. In some states, electricity is several cents cheaper than in others; this is determined by the type and cost of fuel and the type of power plant used to generate the electricity, how the plant is financed and managed, and the distance that the electricity has to travel to reach consumers.

Average cost of electricity per kWh (in cents) in June 2009 and June 2010 for U.S. residences by region and state.

Census Division and State	June 10	June 09
New England	**16.3**	**18.19**
Connecticut	19.47	21.07
Maine	15.37	15.26
Massachusetts	14.62	18.03
New Hampshire	16.19	16.62
Rhode Island	16.53	16.2
Vermont	15.71	15.13
Middle Atlantic	**16.29**	**15.74**
New Jersey	16.86	16.91
New York	19.12	18.51
Pennsylvania	13.33	12.45
East North Central	**11.89**	**11.37**
Illinois	12.6	11.39
Indiana	9.19	9.59
Michigan	12.87	12.56
Ohio	12	11.32
Wisconsin	12.83	12.46
West North Central	**10.27**	**10.02**
Iowa	10.56	10.58
Kansas	10.26	10.08
Minnesota	10.68	10.4
Missouri	10.13	9.79
Nebraska	9.98	9.83
North Dakota	9.38	8.87
South Dakota	9.76	9.26
South Atlantic	**11.3**	**11.45**
Delaware	14.51	14.88
District of Columbia	14.33	13.91
Florida	11.65	12.14
Georgia	10.86	10.72
Maryland	15.22	15.95
North Carolina	10.13	9.9
South Carolina	10.58	10.31
Virginia	10.77	11.15
West Virginia	8.62	7.95
East South Central	**9.73**	**9.88**
Alabama	10.84	11.1
Kentucky	8.38	8.43
Mississippi	10.21	10.46
Tennessee	9.45	9.48
West South Central	**11.12**	**11.42**
Arkansas	9.56	9.41
Louisiana	8.93	7.76
Oklahoma	9.36	8.33
Texas	12.13	12.96
Mountain	**11.26**	**10.61**
Arizona	11.75	11.28
Colorado	11.95	9.95
Idaho	8.21	8.3
Montana	9.34	9.4
Nevada	12.42	12.02
New Mexico	11.34	10.38
Utah	9.16	9.06
Wyoming	9.08	9.23
Pacific Contiguous	**13**	**12.78**
California	15.51	15.02
Oregon	9.13	9.01
Washington	8.26	7.9
Pacific Noncontiguous	**24.18**	**20.62**
Alaska	16.92	17.72
Hawaii	28.36	22.2
U.S. Total	**11.92**	**11.85**

TIP: It takes almost 3kWh of energy to deliver 1 kWh of electricity to you.

Here is another good reason to go solar: according to the DOE, in 2005, an average power plant required 10,210 Btu of energy, the equivalent of almost 3kWh, to generate and deliver 1 kWh of electricity to an end user. In other words, we are paying a high price in energy for our electricity.

How you use electricity in your home

If you calculate the cost of your electricity per kWh month by month, you will probably find that the cost is higher during the months when you use greater amounts of electricity for heating or air conditioning. Your electricity consumption during the months when you are not running an air conditioner or a heater is called your "baseline usage." By comparing your baseline usage to your consumption during winter and summer months you can see how much electricity you are using for heating and cooling.

You can also divide your total annual amount of kWh by 365 (or 366) days to see how your average daily usage compares to the American average of 20 kWh per household.

In the United States and Canada, homes are wired for appliances that require voltages of 110V or 220V. Most household appliances use 110V, but appliances that generate lots of heat such as clothes dryers and kitchen stoves use 220V. Because too much electricity is lost when low voltages travel through cables for long distances, the 12,000V to 25,000V coming from generators in a power plant passes through a step-up transformer that increases the voltage to about 138,000V. The electricity then travels through high-voltage transmission lines to local stations and substations with step-down transformers that deliver the electricity to homes as 220V.

The electricity enters your home through three wires, two carrying 110V each and one neutral ground wire. Every house has a main power panel that distributes electricity to various parts of the house. If you look inside your home's fuse box, you will see that each circuit breaker is labeled for a specific appliance or specific part of the house. The wires going from the panel to most of your power outlets connect one of the 110V wires with the neutral wire to form a circuit that delivers 110V. The wires going to outlets for your clothes dryer and stove connect both 110V wires with the neutral wire to deliver 220V.

The chart below shows a breakdown of how electricity was used in an average American home in 2009. By applying these percentages to your total electricity bill for last year, you can get a rough estimate of how much you paid to operate the various appliances in your home. As you can see, lighting makes up 15 percent of your total electricity consumption. You can start saving money right away by switching to CFL light bulbs

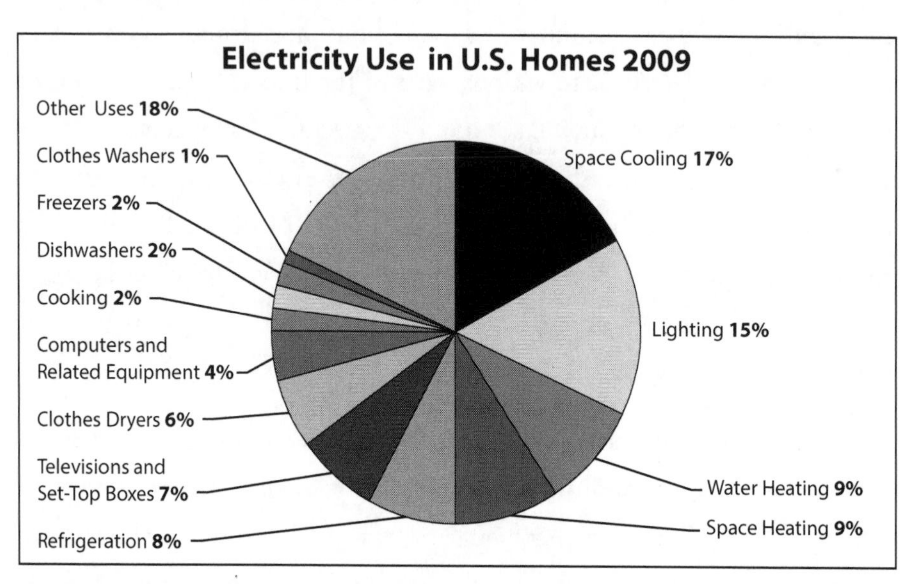

This chart shows the average use of electricity in a 1,624 sq. ft. home in the U.S. in 2009

Looking at how much electricity you use for various purposes around the house can give you an idea of how much you can reduce your electricity bill by switching to solar energy for one or more of these functions, or by upgrading to more energy-efficient appliances. For example, installing a solar water heater could reduce your annual electricity bill by 9 percent.

Today the majority of solar systems are grid-tie systems that draw back-up electricity from your utility when your solar panels are not producing enough, and "sell" excess electricity to your utility when your solar panels produce more power than you need. The climate where you live will have a significant impact on the energy savings you can achieve with a solar system. According to the chart above, space cooling makes up 17 percent of the average household's consumption of electricity. If you live in a hot climate where you use air conditioning 24 hours a day for several months during the summer, a grid-tie system could significantly reduce your utility bills because it would be operating at maximum capacity during the hours when air conditioning is needed most. (Remember that heat can diminish the efficiency of PV panels, so maximum production can be sustained only if the panels are kept below a certain temperature.) During winter months, however, heat is often needed most when skies are cloudy and hours of daylight are short.

The calculations you have made using your electricity bills and average usage data are only rough estimates of how much you can save on utility bills by installing a solar system. Your local utility might have an interactive energy use calculator on its website. You need this information to design a PV system that meets your energy needs.

Cost of electricity provided by a PV system

Buying a PV system to supply your home with electricity is something like prepaying your electricity bill for the next 20 or 25 years at a rate more than double what you are paying for a kWh today. From a purely financial viewpoint, this may not make economic sense. Chapter 1 pointed out that the payback from installing a solar system is not necessarily financial: You have the satisfaction of knowing that you are using clean energy, reducing emission of greenhouse gases into the atmosphere, and lessening your dependence on fossil fuels. You are pioneering a technology that may soon become a mandatory way of life in many parts of the world.

There are many factors to consider when you assess the cost of a solar energy system:

Financing. You will probably use some kind of financing to pay for the initial installation. The interest you will pay on your loan should be factored into the cost of your system, and your monthly payment must be included in your household budget. Your monthly electricity bill will become less or even disappear, depending on the type of system you install, but you will have a new payment to make every month.

Repairs and maintenance. Most solar panels are under warranty for 20 to 25 years, and the other components also come with warranties, but batteries must be replaced every six to eight years. The efficiency of your system may decrease over time as your solar panels age and connections deteriorate. If your system was not installed correctly this deterioration may occur sooner, and your warranties might be invalidated.

Decreasing prices. The cost of solar electricity per Watt Peak has been steadily declining about 4 percent each year. Many governments are actively trying to lower the cost of solar electricity by subsidizing research and development. If you wait a year or two, you may be able to purchase a newer technology at a lower price.

Rebates and tax incentives. The "sticker price" for your PV system could be reduced by as much as 50 percent when you take advantage of all the rebates and tax incentives offered by local utilities and state and federal government. A solar contractor will usually include this information in a price quote.

Increase in the resale value of your home. Retrofitting your home with a solar power system can increase its resale value by as much as $10,000. A home with built-in solar power sells for considerably more than similar non-solar homes in the same market. If you live in an area where solar power is particularly popular, the increase may be even higher.

Future price of electricity from the grid. The cost of electricity from the grid is based on the price of the fuel used to produce it. If unforeseen political or geophysical events suddenly raise the price of coal or oil, utility prices could double within a few years, making your solar power system a much better financial investment.

Solarbuzz, an international solar energy research and consulting company, posts a monthly global solar energy price index on its website (**www.solarbuzz.com/solarindices.htm**). It surveys solar power installation companies all over the world. In June 2010, the average price for a standard grid-tie 2-kilowatt peak system with a battery backup, mounted on the roof of an existing home, was $16,400. This price included full system integration and installation costs. It allowed for a 5 percent interest charge on financing, but did not deduct rebates or tax incentives. At this price, the cost of electricity per kWh in a sunny climate (such as the sun belt of the United States, Africa, Australia, and Asia) was 24.38 cents and in a cloudy climate (such as northern Japan or Germany), 75.41 cents.

A 1-kilowatt peak small-scale solar energy system that provides only a fraction of an average home's energy requirements costs $10,000-$15,000. A 5 kW system that can provide all the energy requirements of most conventional homes costs between $35,000 and $45,000.

The U.S. Department of Energy Solar Decathlon features energy-efficient, solar powered houses built by 20 university teams from North American and Europe.

LEFT: House of Technische Universitat Darmstadt, Germany. Team Germany won first place in the 2009 Decathlon.
CENTER: The University of Illinios at Urbana-Champaign House at the U.S. Doe Solar Decathlon.
RIGHT: Cornell University earned second place in the 2005 Solar Decathlon.

New Home or Existing Home?

The majority of solar installations sold today are retrofitted — they are being installed on existing homes, usually on the roofs, and connected to the existing wiring. Sometimes roof supports have to be augmented to support the extra weight of a solar array or solar water heater, and adjustments have to be made to redirect water runoff from the solar panels when it rains.

TIP: Home ownership and PV systems

A solar PV installation represents a significant investment in your home. Your future plans for your home should be considered when you are deciding what kind of system you want and how much you want to spend. If you own the home or your mortgage is almost paid off, a PV system will increase your home's value and cut down on your monthly utility bills. It can also increase the appeal of your home to a potential buyer. If you still owe a lot on your mortgage and are uncertain about your future in the home, weigh the benefits and risks carefully when making a decision.

It is typically less expensive to install a system during the construction of a new home than to retrofit an existing one. The solar arrays can be incorporated in the architectural design of the new home and the home's energy efficiency can be maximized. Special windows, roof tiles, and other architectural features can serve as solar panels. The house can be designed so solar panels are placed where they receive the most exposure to sunlight, adequate ventilation, and protection from hazardous winds. If you want solar electricity in your new home, your solar contractor should be included in the design process from the beginning.

Converting an older, non-energy efficient home to use solar power for heating, hot water, and electricity can cost $100,000 to $200,000.

Home with solar panels built into the roof.

Conclusion

Using a PV system to provide some or all of your electricity is more practical in some situations than in others. You will get a greater return on your investment if you live in a sunny climate or in a place where energy costs are very high. Solar power is a practical solution for remote locations that are not already connected to the power grid. No matter where you live, you can use some kind of solar application to replace at least a portion of the electricity you buy from a utility.

Most people agree that solar radiation is an excellent source of clean energy and that solar technology will play a vital role in mitigating the environmental crisis brought on by climate change and overpopulation. Governments and utility companies are encouraging as many individuals as possible to invest in grid-tie PV systems, and offer many incentives to those who do.

Your personal decision to go solar depends on many factors besides your geographical location: your personal convictions, your financial resources, the availability and cost of solar products and solar contractors in your neighborhood, your future plans, and the specific physical circumstances of your home.

CHAPTER 4

Making Your Home Energy Efficient

Planning for a solar energy installation automatically makes you think about ways to reduce energy consumption in your home. During your energy use evaluation, you found that the greater part of your electricity bill pays for heating and air conditioning, and for appliances such as dishwashers and clothes dryers. If you can reduce the amount of energy required to heat or cool your home and run appliances, you can reduce the size (and therefore the cost) of the PV system needed to provide that energy. Before proceeding any farther with your solar project, take steps to ensure that your home is using energy as efficiently as possible. Every dollar that you spend on making your home more energy efficient will save you $3 to $5 on the final cost of your PV system.

Home Energy Audit

A professional home energy audit will help you to identify areas in your home where energy is being wasted or lost every day. At the end of the audit you will receive a list of recommended actions to reduce your energy consumption.

Who performs the audit?

Home energy auditors are trained in energy conservation. Several programs train and certify energy auditors, but a home energy audit does not need to be performed by someone with the professional title of "energy auditor." Contractors, energy inspectors, energy raters, and building analysts all perform energy audits. An energy audit is a building inspection that determines how to reduce consumption of energy in a home and save the homeowners money. All energy auditors should be knowledgeable about energy, insulation, heating and cooling systems, water heating systems, and windows and doors.

There are two nationally recognized certifications for energy auditors: the Building Performance Institute (BPI) and the Residential Energy Services Network (RESNET) HERS (Home Energy Rating System) Rater. To become BPI certified, energy auditors must pass a written test and a field test. An HERS Rater must pass a written test and complete five ratings within a year of passing the test. HERS Raters must also take continuing education courses throughout the year. Currently, BPI focuses on older, existing homes, while RESNET focuses on new construction and newer homes.

Finding an auditor

Many utility companies conduct free or discounted home audits to encourage their customers to become more energy efficient. Some of these audits are visual inspections rather than extensive energy audits. Ask what the inspection encompasses. If it does not include blower door and thermography, the inspection might not uncover all of the energy inefficiencies in your home.

If your utility company does not offer a home energy audit, it may be able to recommend a company that does these audits. You can also find an energy auditor in the local Yellow Pages, on the Internet, or through the Better Business Bureau or Chamber of Commerce.

Some solar contractors also do home energy audits. Hiring a solar contractor to do an energy audit does not obligate you to hire him or her to install your PV system. Some companies may offer a discounted rate if you use them for both the audit and the installation.

Auditing your home

A professional energy auditor uses certain tools and procedures to search for energy inefficiencies in your home. A blower door measures air leaks in your building that allow cooled or heated air to escape. A powerful fan mounted into the frame of an exterior door pulls air out of the house and lowers the air pressure inside the house. Higher air pressure outside causes air to flow in through all your home's unsealed openings. You can detect where the air is flowing in by walking around your home and placing your hand over different areas in each room. For example, air will flow in through electrical sockets that are not insulated. The blower door test determines how drafty your house is and identifies where specific leaks are. Repairing these leaks will stop energy wastage by keeping heated or cooled air inside your home. Areas that commonly need repairs include electrical outlets, windows, and doors.

Another procedure used by home energy auditors to detect air leakage is thermography, or infrared scanning. Thermography measures surface temperatures using infrared video and still cameras. Thermography done in cold weather will reveal variations in the temperature of your building's skin, or exterior walls, to show where your home might need additional insulation.

A thorough professional energy audit can take several hours and cost between $300 and $500, depending on the size of your home. It could cost as little as $125 if you get a special discounted rate from a solar contractor or utility company. Many auditors will provide a printed report with detailed recommendations for improving the energy efficiency of your home.

Recommendations for improving energy efficiency

An energy auditor may recommend a number of measures you should take to improve the energy efficiency of your home, including:

Using a vapor barrier or vapor diffusion retarder

A vapor barrier, or vapor diffusion retarder (VDR), is a membrane, coating, or sheet of solid material that impedes vapor diffusion in basements, ceilings, crawl spaces, floors, foundations, and walls. A VDR helps to retain heat in a home and prevents walls, wood, and insulation from coming in contact with moisture. VDRs help control moisture in walls, floors, crawl spaces, basements, and ceilings.

Many new homes are built with VDRs, but some older homes may not have these installed. If your home already has a VDR underneath its siding, you will be able to see it by removing a section of your aluminum siding or other outer layer of your home. If your home has a VDR, then your building envelope, or outer structure, is tight. It is difficult to add a VDR to an existing home. When you are planning a remodeling project or an addition to your home, a VDR should be included.

VDRs are not needed in all climates. According to the DOE, the number of Heating Degree Days in an area determine whether and how to use a VDR. A Heating Degree Day is a measure of how often outdoor daily dry-bulb temperatures fall below an assumed base, normally 65°F (18°C). In climates with cold winters, a VDR should be placed on the interior side of walls, but in hot and humid climates (such as in Florida and Louisiana) a VDR should be added to exterior walls. In some Southern climates homeowners should not use a VDR.

This chart on the DOE website (**www.energysavers.gov/your_home/insulation_airsealing/index.cfm/mytopic=11810**) shows where and how a VDR should be installed on your home. A perm at 73.4°F (23°C) is a measure of the number of grains of water vapor passing through a square foot of material per hour at a differential vapor pressure equal to one inch of mercury (1" W.C.). Any material with a perm rating of less than 1.0 is considered a vapor retarder.

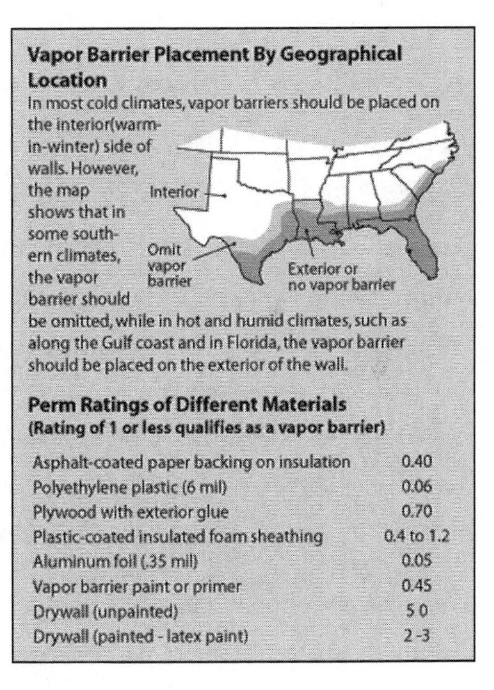

Perm Ratings of Different Materials
(Rating of 1 or less qualifies as a vapor barrier)

Material	Rating
Asphalt-coated paper backing on insulation	0.40
Polyethylene plastic (6 mil)	0.06
Plywood with exterior glue	0.70
Plastic-coated insulated foam sheathing	0.4 to 1.2
Aluminum foil (.35 mil)	0.05
Vapor barrier paint or primer	0.45
Drywall (unpainted)	50
Drywall (painted - latex paint)	2-3

Exterior caulking

Caulking the exterior of your home will seal it off from outside air and moisture. Caulking involves using an impermeable substance to fill in cracks, holes, and seams and make them watertight. On the outside of the building, use a paintable acrylic latex caulk to seal vertical cracks around window and door frames and corner boards. Pay special attention to holes around pipes, electrical outlets, and HVAC (Heating, Ventilation, and Air Conditioning) equipment.

The Home Star Energy Retrofit Act of 2010, informally known as the "cash for caulkers" bill and approved by the House of Representatives in June 2010, will give homeowners up to $3,000 in tax credits to pay for weatherization or insulation improvements.

Buying a quality thermostat

Programmable thermostats allow the temperature in a home to be set to four settings for when you are asleep, away, or at work, and also allow you to warm or cool the home before you return. These thermostats are easy to use and help you to control your heating and cooling costs.

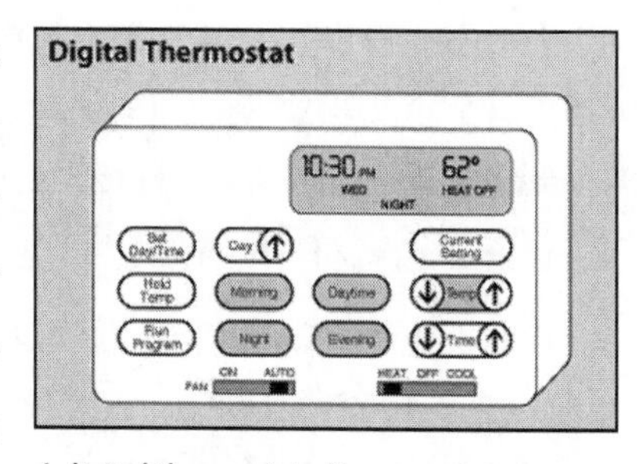

A digital thermostat allows you to manage the temperature in your home when you are asleep or away.

If you have a manual thermostat, you can save by remembering to adjust the temperature when you will not be home for several hours or several days. This will prevent your thermostat from running your heating or cooling system on when no one is home.

Saving money with your thermostat

Excerpt from DOE Energy Savers: Thermostats and Control Systems (www.energysavers.gov/your_home/space_heating_cooling/index.cfm/mytopic=12720)

You can easily save energy in the winter by setting the thermostat to 68°F while you are awake and setting it lower while you are asleep or away from home. By turning your thermostat back 10° to 15° for eight hours, you can save about 5 to 15 percent a year on your heating bill — a savings of as much as 1 percent for each degree if the setback period is eight hours long. The percentage of savings from setback is greater for buildings in milder climates than for those in more severe climates.

In the summer, you can follow the same strategy with central air conditioning, too, by keeping your house warmer than normal when you are away, and lowering the thermostat setting to 78°F (26°C) only when you are at home and need cooling. Although thermostats can be adjusted manually, programmable thermostats will avoid any discomfort by returning temperatures to normal as you wake or return home.

A common misconception associated with thermostats is that a furnace works harder than normal to warm the space back to a comfortable temperature after the thermostat has been set back, resulting in little or no savings. This misconception has been dispelled by years of research and numerous studies. The fuel required to reheat a building to a comfortable temperature is roughly equal to the fuel saved as the building drops to the lower temperature. You save fuel between the time that the temperature stabilizes at the lower level and the next time heat is needed. So, the longer your house remains at the lower temperature, the more energy you save.

Programmable thermostats are generally not recommended for homes with heat pumps, electric resistance heating, steam heat, and radiant floor heating. Electric resistance systems, such as electric baseboard heating, require thermostats capable of directly controlling 120-volt or 240-volt circuits, which are manufactured by only a few companies. A programmable thermostat can cause a heat pump to operate inefficiently during heating mode. Steam heating and radiant floor heating systems take several hours to respond to a thermostat. Recently, however, some companies have begun selling specially designed programmable thermostats that compensate for these limitations.

TIP: How to program your thermostat to save energy.

Note the times when you normally go to sleep and wake up, and the times when everyone in your family is out of the house for four hours or more. Program your thermostat to maintain a temperature that is several degrees warmer in summer and several degrees cooler in winter while you are asleep or away. If you prefer to sleep at a cooler temperature during the winter, you can start the temperature setback a bit ahead of the time you actually go to bed; you probably will not notice the house cooling off as you prepare for bed.

Installing dual pane windows

If you have single-pane windows, upgrading to dual pane windows can make a big difference in energy savings. Energy-efficient windows reduce condensation, reduce fading of draperies and furniture, and enhance the comfort of your home. They can also save you hundreds of dollars in heating and cooling bills each year. The current $1,500 federal tax credit for energy-efficient appliances and building components includes skylights and exterior windows.

The Efficient Windows Collaborative website (**www.efficientwindows.org**) has an interactive program to help you select the most energy-efficient windows for your home. It also provides a printable step-by-step fact sheet for selecting energy efficient windows in each state, and a state-by-state fact sheet for compliance with the International Energy Conservation Code (IECC), the most commonly adopted model energy code for residential buildings.

TIP: Proper installation of windows is essential.

Always follow manufacturers' guidelines for installing windows and use trained professionals. Proper installation ensures an airtight fit, avoids water leakage, and guarantees optimal window performance.

Properly insulating ducts

Houses with forced-air heating and cooling systems typically lose about 20 percent of the air moving through the duct system that distributes heated or cooled air throughout the house. Ducts that are concealed inside walls or between floors are difficult to repair. Exposed ductwork in attics, basements, crawl spaces, and garages can easily be inspected for leaks and poorly sealed connections. Leaks can be sealed with metal tape or duct sealant (also called duct mastic).

Exposed ducts running through attics, garages, or crawlspaces that get very hot in summer or very cold in winter need to be insulated so energy is not lost through the walls of the ducts. If your ducts are not adequately insulated, you can wrap them with a foil-backed insulation (duct wrap).

You can find more information about sealing and insulating ducts online in "*An Introduction to Residential Duct Systems*" (**http://ducts.lbl.gov**), a publication of the Ernest Orlando Lawrence Berkeley National Laboratory.

Weatherstripping your home

Weatherstripping is the application of a narrow strip of material to fill in small cracks around windows and doors. According to home improvement store Lowe's, energy loss from a tiny space between an exterior door and its threshold is equal to energy loss from a two-inch hole in the wall. Lowe's estimates that closing those gaps can save you up to 15 percent in heating and cooling costs. DOE experts agree that caulking and weatherstripping will pay for themselves in energy savings within one year.

Weatherstripping

Excerpt from U.S. Department of Energy — Energy Efficiency and Renewable Energy: Energy Savers (www.energysavers.gov/your_home/insulation_airsealing/index.cfm/mytopic=11280)

You can use weatherstripping in your home to seal air leaks around movable joints, such as windows or doors. Before applying weatherstripping in an existing home, you need to locate air leaks and assess your ventilation needs. Unless properly ventilated, an airtight home can seal in indoor air pollutants. Ventilation also helps control moisture, which can fuel the growth of mold and mildew.

- To determine how much weatherstripping you will need, add the perimeters of all windows and doors to be weatherstripped, then add 5 to 10 percent to accommodate any waste. Also consider that weatherstripping comes in varying depths and widths.

* Choose a type of weatherstripping that will withstand the friction, weather, temperature changes, and wear and tear associated with its location. For example, when applied to a door bottom or threshold, weatherstripping could drag on carpet or erode as a result of foot traffic. Weatherstripping in a window sash must accommodate the sliding of panes — up and down, sideways, or out. The weatherstripping you choose should seal well when the door or window is closed while allowing it to open freely.
* Choose a product for each specific location. Felt and open-cell foams tend to be inexpensive, susceptible to weather, visible, and inefficient at blocking airflow. However, the ease of applying these materials may make them valuable in low-traffic areas. Vinyl, which is slightly more expensive, holds up well and resists moisture. Metals (bronze, copper, stainless steel, and aluminum) last for years and are affordable. Metal weatherstripping can also provide a nice touch to older homes where vinyl might seem out of place. Take durability into account when comparing costs.
* You can use more than one type of weatherstripping to seal an irregularly shaped space.

Properly insulating walls and attics

According to the DOE, only 20 percent of homes built before 1980 are well-insulated. A good insulating system includes a combination of products and construction techniques that protect a home from outside hot or cold temperatures, protect it against air leaks, and control moisture.

Insulation is measured in R-values. The higher the R-value of your insulation, the better your walls and roof will resist the transfer of heat.

Insulation is made of a variety of materials. There are four basic types:

Rolls and batts are flexible blankets made from mineral fibers and are sold in widths adapted to standard spacing of wall studs and attic or floor joists.

Loose-fill insulation, usually loose fibers or fiber pellets made of fiberglass, rock wool, or cellulose, is blown into spaces using special pneumatic equipment. It conforms easily to building cavities and attics.

Rigid foam insulation is more expensive than fiber insulation but is effective in buildings with space limitations. Foam insulation R-values range from R-4 to R-6.5 per inch of thickness, up to two times greater than most other insulating materials of the same thickness.

Foam-in-place insulation is foam that can be blown into walls. It can reduce air leakage, if blown into cracks around window and door frames.

Do-it-yourself energy audit

You do not have to hire a professional to perform an energy audit for you. Going through your home looking for inefficiencies might make an interesting weekend project. The U.S. Department of Energy (DOE) (**www.energysavers.gov**) offers step-by-step instructions on its website for conducting your own energy audit. You might decide to call a professional energy auditor after you have performed your own audit.

Do-It-Yourself Home Energy Assessments

Excerpt from the U.S. Department of Energy — Energy Efficiency and Renewable Energy: Energy Savers (www.energysavers.gov/your_home/energy_audits/index.cfm/mytopic=11170)

You can easily conduct a do-it-yourself home energy assessment (also known as a home energy audit). With a simple but diligent walk-through, you can spot many problems in any type of house. When assessing your home, keep a checklist of areas you have inspected and problems you found. This list will help you prioritize your energy efficiency upgrades.

Locating Air Leaks

* First, make a list of obvious air leaks (drafts). The potential energy savings from reducing drafts in a home may range from 5 to 30 percent per year, and the home is generally much more comfortable afterward. Check for indoor air leaks, such as gaps along the baseboard or edge of the flooring and at junctures of the walls and ceiling. Check to see if air can flow through these places:
    * Electrical outlets
    * Switch plates
    * Window frames
    * Baseboards
    * Weatherstripping around doors
    * Fireplace dampers
    * Attic hatches
    * Wall- or window-mounted air conditioners

* Also look for gaps around pipes and wires, electrical outlets, foundation seals, and mail slots. Check to see if the caulking and weatherstripping are applied properly, leaving no gaps or cracks, and are in good condition.

* Inspect windows and doors for air leaks. See if you can rattle them — movement means possible air leaks. If you can see daylight around a door or window frame, then the door or window leaks. You can usually seal these leaks by caulking or weatherstripping them. Check the storm windows to see if they fit and are not broken. You may also wish to consider replacing your old windows and doors with newer, high-performance ones. If new factory-made doors or windows are too costly, you can install low-cost plastic sheets over the windows.

* If you are having difficulty locating leaks, you may want to conduct a basic building pressurization test:

1. First, close all exterior doors, windows, and fireplace flues.
2. Turn off all combustion appliances such as gas burning furnaces and water heaters.
3. Then turn on all exhaust fans (generally located in the kitchen and bathrooms) or use a large window fan to suck the air out of the rooms.

This test increases infiltration through cracks and leaks, making them easier to detect. You can use incense sticks or your damp hand to locate these leaks. If you use incense sticks, moving air will cause the smoke to waver, and if you use your damp hand, any drafts will feel cool to your hand.

- On the outside of your house, inspect all areas where two different building materials meet, including:
    - All exterior corners
    - Where siding and chimneys meet
    - Areas where the foundation and the bottom of exterior brick or siding meet.
- You should plug and caulk holes or penetrations for faucets, pipes, electric outlets, and wiring. Look for cracks and holes in the mortar, foundation, and siding, and seal them with the appropriate material. Check the exterior caulking around doors and windows, and see whether exterior storm doors and primary doors seal tightly.
- When sealing any home, you must always be aware of the danger of indoor air pollution and combustion appliance "backdrafts." Backdrafting is when the various combustion appliances and exhaust fans in the home compete for air. An exhaust fan may pull the combustion gases back into the living space. This can obviously create a very dangerous and unhealthy situation in the home.
- In homes where a fuel is burned (i.e., natural gas, fuel oil, propane, or wood) for heating, be certain the appliance has an adequate air supply. Generally, one square inch of vent opening is required for each 1,000 Btu of appliance input heat. When in doubt, contact your local utility company, energy professional, or ventilation contractor.

Insulation

Heat loss through the ceiling and walls in your home could be very large if the insulation levels are less than the recommended minimum. When your house was built, the builder likely installed the amount of insulation recommended at that time. Given today's energy prices (and future prices that will probably be higher), the level of insulation might be inadequate, especially if you have an older home.

- If the attic hatch is located above a conditioned space, check to see if it is at least as heavily insulated as the attic, is weatherstripped, and closes tightly. In the attic, determine whether openings for items such as pipes, ductwork, and chimneys are sealed. Seal any gaps with an expanding foam caulk or some other permanent sealant.
- While you are inspecting the attic, check to see if there is a vapor barrier under the attic insulation. The vapor barrier might be tarpaper, Kraft paper attached to fiberglass batts, or a plastic sheet. If there does not appear to be a vapor barrier, you might consider painting the interior ceilings with vapor barrier paint. This reduces the amount of water vapor that can pass through the ceiling. Large amounts of moisture can reduce the effectiveness of insulation and promote structural damage.

Make sure that the attic vents are not blocked by insulation. You also should seal any electrical boxes in the ceiling with flexible caulk (from the living room side or attic side) and cover the entire attic floor with at least the current recommended amount of insulation.

* Checking a wall's insulation level is more difficult. Select an exterior wall and turn off the circuit breaker or unscrew the fuse for any outlets in the wall. Be sure to test the outlets to make certain that they are not "hot."

* Check the outlet by plugging in a functioning lamp or portable radio. Once you are sure your outlets are not getting any electricity, remove the cover plate from one of the outlets and gently probe into the wall with a thin, long stick or screwdriver. If you encounter a slight resistance, you have some insulation there. You could also make a small hole in a closet, behind a couch, or in some other unobtrusive place to see what, if anything, the wall cavity is filled with. Ideally, the wall cavity should be totally filled with some form of insulation material. Unfortunately, this method cannot tell you if the entire wall is insulated, or if the insulation has settled. Only a thermographic inspection can do this.

* If your basement is unheated, determine whether there is insulation under the living area flooring. In most areas of the country, an R-value of 25 is the recommended minimum level of insulation. The insulation at the top of the foundation wall and first floor perimeter should have an R-value of 19 or greater. If the basement is heated, the foundation walls should be insulated to at least R-19. Your water heater, hot water pipes, and furnace ducts should all be insulated. *For more information, see our insulation section.*

Heating/Cooling Equipment

* Inspect heating and cooling equipment annually, or as recommended by the manufacturer. If you have a forced-air furnace, check your filters and replace them as needed. Generally, you should change them about once every month or two, especially during periods of high usage.

* Have a professional check and clean your equipment once a year.

If the unit is more than 15 years old, you should consider replacing your system with one of the newer, energy-efficient units. A new unit would greatly reduce your energy consumption, especially if the existing equipment is in poor condition.

* Check your ductwork for dirt streaks, especially near seams. These indicate air leaks, and they should be sealed with a duct mastic. Insulate any ducts or pipes that travel through unheated spaces. An insulation R-Value of six is the recommended minimum.

Lighting

* Energy for lighting accounts for about 10 percent of your electric bill. Examine the wattage size of the light bulbs in your house. You may have 100-watt (or larger) bulbs where 60 or 75 watts would do.

* Consider compact fluorescent lamps for areas where lights are on for hours at a time. Your electric utility may offer rebates or other incentives for purchasing energy-efficient lamps.

Steps for Decreasing Energy Use

Even without performing a full-scale energy audit, there are steps you can take to make your house more energy efficient. Before beginning your solar installation, make time for some home improvements that will reduce your energy consumption. If you do this several months ahead of the installation, you will have time to compare your utility bills to previous years and see if your energy requirements have been significantly lowered.

Your attic

Adequate insulation in your attic will keep your home warmer in winter and cooler in summer. The online DOE Zip Code Insulation Calculator (**www.ornl.gov/~roofs/Zip/ZipHome.html**) will give you recommended insulation levels tailored to your home.

Install a simple attic fan to circulate hot air out of your attic during the summer months and help keep your air conditioning bills lower.

Inadequate ventilation in your attic shortens the life of your roof, makes upstairs living spaces hot in the summer, and places heavier demands on your air conditioning system. Ventilation requirements are determined by factors such as roof color (black roofs are hotter than white), the amount of insulation in the attic floor, the existence of a vapor barrier, whether openings in the attic are covered by screens, and the amount of shade protecting the roof. The basic standard is one square foot of cross-ventilation for every 300 feet of attic space. Static ventilators such as fixed gable louvers or roof ridge vents work best in combination with vents located around the base of the attic. For example, soffit vents in the eaves allow cooler air to enter from below and escape as hotter air through the gable or ridge openings in the roof above. In the winter, proper attic ventilation keeps the attic cold and protects your roof from structural damage from ice that forms when snow melts and refreezes on a warm roof. Make sure that vent openings are not obstructed by insulation.

TIP: Adding insulation to your attic is one of the most cost-effective ways to make your home more comfortable year-round.

Measure the thickness of the insulation in your attic. If it is less than R-30 (11 inches of fiberglass or rock wool or 8 inches of cellulose), you could probably benefit by adding more. According to the DOE, most U.S. homes should have between R-30 and R-60 insulation in the attic. Also add insulation to the attic trap or access door.

Windows, doors, and electrical outlets

Check your window and door frames for any small gaps or holes that allow air to escape from or enter your home. If, for instance, some of your windows have tiny gaps measuring 1/10 of an inch square, it would take ten of these gaps to make up a 1-inch hole. Thirty gaps 1/10 inch square equal a 3-inch hole in your wall. Visualize a 3-inch hole in your wall — how much of your air-conditioned or heated air can escape through that hole?

TIP: A candle can help you find leaks

Use a lighted candle or cigarette to check for air leaks around windows and electrical outlets. If the candle flame flickers or the smoke is disturbed, you will know that there is a leak in that area.

Install special insulators under the electrical outlets set in the perimeter walls of the home.

Put new weatherstripping along the bottom of all doors that lead outside. If this is difficult, you can lay long narrow cloth bags or tubes filled with sand along the bottom of doors to block drafts.

TIP: A Do-It-Yourself Guide to Sealing and Insulating with Energy Star®

The Energy Star® website offers a do-it-yourself guide for sealing and insulating your home (**www.energystar.gov/index.cfm?c=diy.diy_index**).

Energy Star® estimates that a knowledgeable homeowner or skilled contractor can save up to 20 percent on heating and cooling costs (or up to 10 percent on their total annual energy bill) by sealing and insulating.

Energy-efficient appliances

You can reduce your energy consumption by replacing old appliances with newer, more energy-efficient models. Energy Star® (**www.energystar.gov**), a joint program of the U.S. Environmental Protection Agency (EPA) and the DOE, gives an Energy Star® rating to appliances that meet certain standards for energy efficiency. To qualify, products must have the features and performance demanded by consumers in addition to increased energy efficiency. If a qualified product is more expensive than a conventional, less-efficient counterpart, those who purchase it must be able to recover their investment in increased energy efficiency through savings on their utility bills within a reasonable period of time. The product's energy consumption and performance must be measurable and verifiable with testing. Products must be clearly labeled with information on energy consumption.

When shopping for a new appliance, look for one with an Energy Star® rating. Each Energy Star® appliance has a label that outlines how much you can expect to save on your utility bill by switching to that energy-efficient model. For example, a refrigerator with new energy-saving features can save as much as 10 percent in annual consumption of electricity. To qualify for an Energy Star® rating, the refrigerator must use 20 percent less energy than a standard fridge. According to the DOE, replacing your old refrigerator from the 1980s with an Energy Star® model will save you more than $100 annually on your utility bills, and you can save $200 by replacing a refrigerator from the 1970s.

Dishwashers with an Energy Star® rating save on both water consumption and electricity. The DOE reports that dishwashers made before 1994 waste about 8 gallons of water per cycle compared to new Energy Star® models. If you replace your old model with one that is Energy Star®-qualified, you can save an extra $40 a year on your utility bills.

According to the Energy Star® website, Americans saved enough energy to power 10 million homes in 2009 alone, helping reduce greenhouse gas emissions by an amount equivalent to the emissions of 12 million cars. In monetary terms, a savings of $6 billion was realized in 2009.

Incentives for energy-efficient appliances

The American Recovery and Reinvestment Act of 2009 included a "Cash for Appliances" rebate for the purchase of new Energy Star®-qualified appliances, including boilers, central air conditioners, washers, dryers, dishwashers, freezers, oil and gas furnaces, refrigerators, room air conditioners, and heat pumps. In 2010, $300 million in federal funds had been allocated to the program. Each state determined the amount consumers would be reimbursed under the plan, ranging from $50 to $250.

Solar-powered appliances

If you are planning an off-grid PV system, particularly for a seasonal vacation cabin or a remote location, you can use appliances that are designed to run on solar power sources. A variety of 12v/24v appliances are on the market. A solar-powered refrigerator is smaller and more expensive, but it avoids the need to purchase extra solar panels and battery storage to provide electricity for a conventional refrigerator. SunDanzer offers a model 46 inches wide by 34.5 inches high that retails for about $1,200. (**www.sundanzer.com/Home.html**).

Solar-powered refrigerator.

Energy-efficient lighting

Only 10 percent of the energy used by an incandescent bulb produces light; the remaining 90 percent is given off as wasted heat energy. Compact fluorescent bulbs (CFLs), in contrast, give off the same amount of light using only ¼ of the electricity. CFLs are narrow twisted glass tubes containing a mixture of three phosphors that give off light when exposed to ultraviolet light from mercury atoms. They are made to fit into standard light bulb sockets and are sometimes covered with globes to resemble incandescent bulbs.

CFL bulbs can cost $4 to $6 each, as opposed to as little as 50 cents for some incandescent bulbs, but the extra expense is recouped in energy savings by the time the bulb has been in use for 500 hours. Incandescent bulbs generally have to be replaced after 500 to 2,000 hours of use, while

CFLs are supposed to last 8,000 hours. The DOE is actively campaigning to encourage homeowners to switch to CFLs.

60 watt incandescent light globe × 8 hours a day = 480 watts

15 watt CFL bulb × 8 hours a day = 120 watts

Compact Fluorescent Lighting used in an exterior light fixture.

A CFL can qualify for an Energy Star® rating if can save more than $40 in electricity costs over its lifetime, uses about 75 percent less energy and lasts up to 10 times longer than standard incandescent bulbs, and produces about 75 percent less heat to cut energy costs associated with home cooling. According to the DOE, if every American home replaced just one light with a light that's earned the Energy Star®, we would save enough energy to light 3 million homes for a year, save about $600 million in annual energy costs, and prevent 9 billion pounds of greenhouse gas emissions per year, equivalent to the emissions from about 800,000 cars.

The DOE recommends replacing the incandescent bulbs that are turned on in your home for the longest periods with similar CFLs. You can find an ENERGY STAR® Choose Your Light guide on the Energy Star® website (**www.energystar.gov/index.cfm?c=cfls.cfls_choose_guide**).

TIP: CFLs contain mercury and must be disposed of properly.

Each CFL contains about 5 mg of mercury, a toxic heavy metal that can cause serious health problems if inhaled or ingested over a period of time or in large enough doses. This is less than the amount of mercury contained in a thermometer and presents no danger to occupants of the home, but old CFLs should be recycled at a hazardous waste facility to prevent them from going into landfills.

If a compact fluorescent lamp breaks in your home, open nearby windows immediately to disperse any mercury vapor that may escape. Carefully sweep up the glass fragments, place them in a sealed plastic bag and dispose of them with your other household trash. Wipe the area with a disposable paper towel to remove any remaining fragments. Do *not* try to pick up glass fragments with your hands, and do *not* use a vacuum.

Burning fossil fuels such as coal to produce electricity releases mercury into the air. Reducing the amount of electricity consumed by using CFLs prevents mercury from being released into the atmosphere.

Light emitting diode (LED) technology, also known as solid state lighting (SSL), is gaining popularity as energy-efficient lighting. According to the DOE, it has the potential to be 10 times more energy-efficient than traditional incandescent lighting. As part of the Recovery Act of 2009, the U.S. government allotted $37 million to help develop LED products. Europe, Japan, China, and South Korea have similar government initiatives. The current price of LEDs restricts their use in private residences, but as more products come to market, the manufacturing cost will go down.

Only a limited number of LED products are Energy Star®-rated, but the DOE expects that number to grow substantially by 2011. Some LEDs experience problems such as flickering, color shifting, dimness, uneven light, and the use of power when turned off, after less than a year of use. The DOE recommends purchasing Energy Star®-rated LEDs that have met performance and efficiency standards. LED products for use in the home include under-cabinet kitchen lights, shelf-mounted lights, desk lamps, recessed lighting, cove lighting, outdoor step and pathway lights, and outdoor decorative lights.

Landscaping for more energy conservation

You can reduce the amount of energy needed to heat and cool your house by the strategic placement of trees and shrubs.

Windbreaks

A windbreak can lower the wind chill near your home during cold winter months. Wind chill occurs when wind speed lowers the outside temperature. For example, if the outside temperature is 10°F (-12°C) and the wind speed is 20 miles per hour (32 kilometers per hour), the wind chill is -24°F (-31°C).

To plant an effective windbreak, you need to know which plants grow best in your region and what direction the wind comes from around your house. Trees and shrubs with low crowns work best because they block wind close to the ground. Trees, bushes, and shrubs are often planted together to create a wind barrier from the ground up. The most common type of windbreak consists of dense evergreen trees and shrubs planted to the north and northwest of a home. Evergreen trees combined with a wall, fence, or manmade earth berm can deflect or lift the wind over the home.

A windbreak will reduce wind speed for a distance of as much as 30 times the windbreak's height. For maximum protection, your windbreak should be planted at a distance from your home of two to five times the mature height of the trees. Be careful not to plant the trees too close to the house where they will block the winter sunlight or cast shade on your solar panels.

Summer winds can have a cooling effect, especially at night. However, if summer winds are hot and you live in a climate where you use air conditioning all summer, you may want to block the winds from circulating near your home.

Insulation

Planting shrubs, bushes, and vines next to your house creates dead air spaces that insulate your home in both winter and summer. Place plants so that there will be at least at least a foot (30 centimeters) of space between them and the wall when they are fully grown.

Shading

Shading and evapotranspiration (the process by which a plant actively moves and releases water vapor) from trees can reduce surrounding air temperatures as much as 9°F (5°C). Because cool air settles near the ground, air temperatures directly under trees can be as much as 25°F (14°C) cooler than air temperatures above nearby blacktop.

To use shade effectively you must know the size, shape, and location of the shadows cast by your trees and plants as the sun moves across the sky during the day. Trees can be found with appropriate sizes, densities, and shapes for almost any shading application. You must also know which trees grow well in your region. Deciduous trees block solar heat in summer but let sunlight through in winter when they lose their leaves. Evergreen trees provide shade year-round. Homes in cooler climates may not need shade at all. Shading an air conditioner can increase its efficiency by as much as 10 percent. Trees, shrubs, vines, and groundcover plants can reduce heat radiation and cool the air by shading the ground and pavement around your home. A patio or driveway can be shaded with a large bush or row of shrubs, a hedge, or vines on a trellis.

Solar energy absorbed through your roof and windows increases the heat inside your home during the summer, but shade on your roof may interfere with the operation of your solar panels. Locate trees where they will not block sunlight falling on your solar array. Even the bare branches of deciduous trees can affect the efficiency of your solar panels.

Deciduous trees with high, spreading crowns to the south of your home will provide maximum shade for your roof during summer. Trees and shrubs planted to the west will protect your home from sun angling toward the windows during the afternoon.

Breaking bad habits

Another way to reduce your energy consumption is to change your household's energy habits. If your family is not already practicing energy conservation, a few adjustments can have an immediate effect on your utility bill:

* Turn electronics off when they are not in use. If a light or television is on when no one is in a room, turn both electronics off.
* Lower your thermostat a few degrees if no one is home. If you raise your thermostat by 1 degree in the summer, your cooling bill can be reduced by 2 percent. In the winter, your heating bills can be reduced by 3 percent if you lower your thermostat by 1 degree. Wear an extra sweater indoors in winter and lighter clothing in the summer.

- Always close exterior doors when operating an air conditioner or heating system.
- Turn off your computer when it is not in use or put it on power save mode.
- Replace your incandescent light bulbs with compact fluorescent bulbs, reducing your electricity usage for lighting by 75 percent.
- Avoid opening the refrigerator as much as possible.
- Use LEDs or battery-powered nightlights in hallways and bathrooms at night instead of incandescent bulbs.
- Unplug appliances that are not in use, such as coffee makers and juicers.
- Unplug your cell phone, MP3 player, and camera chargers when you are not charging.
- Use power strips to switch off televisions, home theater equipment, and stereos when they are not in use. The "standby" consumption of these devices can be equivalent to that of a 75- or 100-watt incandescent bulb running continuously.
- Control your use of hot water. Do your laundry with cold water. Switch to a low-flow showerhead and turn off the shower while you are applying soap and shampoo.

CHAPTER

5

Do-It-Yourself or Hire a Contractor?

In the first chapters of this book, you have learned that installing a solar energy system in your home is more complicated than simply attaching some solar panels to your roof and plugging in. A full-scale PV project may involve installing and hooking up a generator or a bank of batteries, wiring electrical connections, mounting a solar array somewhere on your roof or near your house, and installing an inverter to connect to the power grid. The physical installation is only part of a successful solar project. A certain amount of expertise is needed to anticipate local weather conditions, determine your energy requirements, choose the right equipment, locate a solar array in the optimal position, and provide adequate backup power.

Over time, a PV system experiences a lot of wear-and-tear. Sunlight, rain, and snow are corrosive. The electrical connections must be able to withstand repeated exposure to moisture. Even if a system works well when you first install it, deterioration can diminish its performance within a few years. A solar contractor has the experience and know-how to prevent problems at potential trouble spots.

Even when you have the knowledge and resources to undertake a do-it-yourself project, it may be faster and more cost-effective to hire a contractor to do the work for you. This chapter will help you decide whether you are prepared to install your own solar system, and guide you through the process of finding a reliable contractor if you need one.

Questions to Ask Yourself Before You Begin a Do-It-Yourself Project

The questions below will help you decide whether you have the necessary skills to install your own PV system, and give you an idea of the amount of time and money you will be investing in the project.

What is the size and scope of your proposed PV installation?

Are you installing a couple of solar panels to generate power for a water pump or an isolated storage shed, or are you planning a full off-grid system with batteries or a back-up generator? Will you be connecting your system to the grid? Some smaller solar projects can be bought as complete kits. A large-scale PV installation calls for multiple skills and probably requires the services of a licensed electrician. Every PV installation is unique, designed to supply specific power needs and conform to the physical circumstances of the site.

Why do you want to do this project yourself?

Are you planning on doing this project yourself to save money, or because you want the experience of learning firsthand about solar installations? The cost of purchasing or renting tools and acquiring licenses and permits may cancel out most of the savings you realize by doing the whole project yourself. A poorly installed solar array will fail to produce the power you expect from it; you might have to make several attempts before you get it right.

On the other hand, if part of your motivation for going solar is a desire to educate yourself and others, the extra time it takes you to do research and acquire new skills might be worthwhile.

If you live in a remote location where you will have to do the maintenance and repairs on your system, you will want to learn as much about it as you can. In that case, doing the installation yourself can be a valuable experience.

You can hire a contractor and learn from him or her by participating in the work, or use a contractor for part of the job and do the rest yourself. For example, you could hire an electrician to do the wiring and electrical connections while you install the mounts for your solar array and build a battery box.

Have you worked with electricity before? How much experience do you have?

Solar panels typically come with schematic diagrams of the electric circuits that may be difficult for an inexperienced layman to understand. For a system to run at maximum efficiency, the electrical connections must be robust, and wires of the correct sizes must be used. Working with electricity can be very dangerous during the installation process, and a faulty installation could result in a fire or electrocution hazard that would endanger the lives of your family.

Are you prepared to attend classes to get a certification?

Do you know what licenses and permits are required?

Most localities have building codes and a permitting system to ensure that construction and electrical wiring meet certain standards of safety and quality. Contact the zoning and building enforcement divisions of your local jurisdiction, describe your project, and ask what the process is for getting permits and having inspections done. Your homeowner's association may have additional rules and requirements.

Some common code violations for PV systems are:

- Improper wiring.
- Exceeding roof load.
- Obstructing side yards and setbacks.
- Erecting unlawful PV frames.
- Erecting unlawful protrusions on roofs.
- Installing components too close to a street or lot boundary.

Common code violations for solar water heating systems:

- Unacceptable heat exchangers.
- Unlawful tampering with potable water supplies.

A solar contractor knows how to design a system that complies with local building codes and will do the work of getting permits and scheduling inspections. It is not unusual to make several visits to local zoning offices before a solar installation is complete.

Are you planning to apply for a grant, a rebate, or a loan? Will you be connecting to a local utility?

Many rebates, grants, and loans require that the PV system be installed by a registered solar contractor according to certain specifications.

Warranties and buy-back programs stipulate that your solar system must be installed by a licensed professional. Installing the system yourself would disqualify you and make you responsible for repairing any damaged or malfunctioning equipment at your own expense.

Do you have the necessary tools and equipment?

Any project is much easier if you have the right equipment and the best tools for the job. A solar contractor already has all the tools and supplies for a successful solar installation and knows how to use them. You might find yourself making one trip after another to the home improvement store to buy tools that you will use only once. Tools needed for a solar installation include:

- Multimeter, also known as a multitester or a volt/ohm meter (VOM) — a meter that measures multiple electrical variables, including voltage, current, and resistance.
- Wire strippers
- Wire nuts and electric tape
- Carpenter's square
- Power drill and bits
- Power saw
- Measuring tapes (25-foot and 100-foot)
- Hammers

* Screwdrivers
* Adjustable wrenches and pliers
* Utility knives
* Levels (2-feet and 4-feet)
* Extension cords
* Sturdy wood or insulated fiberglass ladders
* Work gloves, safety glasses, and protective clothing

When you consider the expense of purchasing tools, a contractor might not seem expensive after all!

How much time do you have?

Before you begin a solar project, you will spend a considerable amount of time doing research, comparing prices and products, surveying your site, and getting bids from potential contractors. Depending on the complexity of your PV installation, and whether you have to wait for inspections and approvals, it could take you days and even weeks to finish the project.

Are you physically up to the job?

Do you have a good sense of balance and the stamina to crawl around on a roof dragging heavy solar panels?

What Solar Contractors Do

A solar contractor is qualified and licensed to carry out solar installations. A contractor may work independently or represent a particular brand or product. Solar contractors can:

* Help you evaluate your energy needs and design a solar installation appropriate for your home.
* Make a detailed survey of your site and determine the best locations for the solar array and other components of your system.
* Recommend equipment and help you with pricing and purchasing components.
* Get all necessary local permits and work with inspectors during the approval process.
* Correctly install the components of your system, mount your solar array, build a battery housing, and do all the necessary carpentry and fitting.
* Do the electrical wiring and connections, install an inverter, and connect you to the grid.
* Familiarize you with rebates, tax incentives, local energy programs, and financing options for your solar system.
* Instruct you about the care and maintenance of your new system
* Provide technical support and warranty services.

Shopping for a Solar Contractor

The best place to begin looking for a solar contractor is through referrals from friends or acquaintances who have had solar systems installed. If you see solar panels on the roof of a house in your neighborhood as you drive by, stop and ask who did the installation and what the homeowner's experience was like.

Look for solar professionals listed in your local phone directory or Yellow Pages (**www.yellowpages.com**). You can also search the membership directories of your local Better Business Bureau (**www.bbb.org**) or Chamber of Commerce (**www.uschamber.com/chambers/directory/default**). On the website of the North American Board of Certified Energy Practitioners (NABCEP), a volunteer board made up of representatives of the solar industry, NABCEP certificants, renewable energy organizations, state policy makers, and educational institutions, that offers a certification program for solar professionals, you can search a database of certified solar installers (**www.nabcep.org/installer-locator**). Energy Matters LLC, a developer of technologies for the solar market, has a searchable database of 3,720 solar, wind, and renewable energy professionals, accessible at Find-Solar.org (**www.find-solar.org**) and Solar Estimate.org (**www.solar-estimate.org**).

If your state has a solar energy program, you may be able to obtain a directory of solar energy companies. Some manufacturers of solar panels, such as Kyocera (**www.kyocerasolar.com**) and Uni-solar (**www.uni-solar.com**), have lists of their dealers and contractors on their websites. You could also contact a manufacturer directly and ask for a referral. Home improvement stores, realtors, and even the lender from whom you are getting financing for your project might also be able to recommend local solar contractors.

Questions to ask a potential contractor

Once you have compiled a list of local solar contractors, you should request a quote from at least three of them. There can be a great variation in products and services offered, and in cost. Interviewing three contractors will give you a good sense of what your project involves and enable you to select someone with whom you can work well.

Meet each contractor face-to-face at least once. A personal meeting will give you an opportunity to see whether you will be able to communicate easily with him or her. If you are planning to install a PV system in a home that you are building, try to visit the contractor at a job site, so you can see how the work there is organized. You can also ask the contractor to show you examples of systems similar to the one you will be installing.

Here is a list of questions to ask a potential solar contractor:

* Are you licensed? What licenses do you have? Are your workers licensed? Are you certified for solar installations?
* What is your contractor's license number?
* Are you insured and bonded?
* What kind of experience have you had in solar energy installation? How long has your company been in business?
* How many solar installations have you done in the past two years?
* Do you do all the work yourself? Do you hire subcontractors? Are they licensed?
* Do you handle the permits and inspections I need for the installation?
* What is your typical procedure for planning and carrying out a project?
* Can you help me with incentives and rebates offered by my utility company and the state and federal government?

* Can you recommend a lender that you have worked with before to finance this project?
* Are you a dealer for a particular manufacturer of solar products? Why would you recommend this manufacturer over other solar products?
* What is your typical payment schedule?
* What is your policy on warranties and after-sales service?
* Can you provide references?

Check with your local Better Business Bureau (BBB) and your state consumer affairs department to see if any complaints have been filed against a particular contractor. This can usually be done online. Investigate how any complaints were resolved and whether the problem was related to the quality of the work. A complaint does not mean you should automatically dismiss that contractor, but it is a red flag. Ask the contractor to explain his or her side of the story.

TIP: Google each contractor's name.

Type each contractor's name in a search engine and see what kind of results turn up. You will probably find only business information. If your search pulls up notices of legal action against a company, blog entries complaining about a contractor's work or business ethics, or news articles pertaining to complaints, you should seriously consider using a different contractor.

Contact the references and ask about their experience working with the contractor. Find out whether their solar project was similar to yours, how well it is working out, and whether they are satisfied with the contractor's customer service.

Licenses and certifications

A license is an official document issued by a state or local government entity allowing a person to carry out a particular type of business activity or trade. It ensures that the person holding the license has the necessary skills and education to carry out work in compliance with state standards of safety and quality. To get a license, a tradesman must typically pass an examination. Most certifications are voluntary. A third-party organization verifies that a contractor has followed an education program and met all the requirements to be certified. A certification indicates that a contractor has made extra effort to be recognized as a specialist in his field.

Licenses

Licensing requirements vary from state to state. According to the Database of State Incentives for Renewables & Efficiency (DSIRE) (**www.dsireusa.org**), only 14 states plus Puerto Rico had solar contractor licensing requirements as of September 2010. They are California, Oregon, Utah, Nevada, Arizona, Hawaii, Florida, Arkansas, Michigan, Connecticut, Louisiana, Tennessee, Virginia, and Vermont. Solar contractor licensing requirements will probably change as the industry continues to grow. Though your state may not currently require a solar contractor license, confirm that the contractor you hire to work on your home has the necessary background to perform the work. For example, if you are having a solar water heater installed, you should check whether your contractor has a plumber's license. If you are installing a photovoltaic system, your contractor should have an electrical license. If your contractor does not have the necessary background and does not install your system correctly, the result may be poor performance or a safety hazard.

Some states that do not require licensing leave that matter to the local jurisdictions, and not all jurisdictions may require licenses. Permits for solar panel installations, however, require certain licenses for those who are performing the work. A solar product manufacturer may refuse to honor a buy-back program or a warranty if the product was not installed by a licensed contractor. Read the details of your warranty carefully.

According to DSIRE, contractors without full electrical and plumbing licenses can get a solar specialty license if they only plan to install solar systems. Ask your contractor if he or she, or any of the contractor's employees has an electrical or plumbing license (depending on if you are having PV or solar water heater installation), or a solar specialty license. You should not hire an unlicensed contractor to do a sophisticated PV installation. Licensing may not be a problem if your solar project is not subject to permits.

Certifications

The nationally recognized North American Board of Certified Energy Practitioners (NABCEP) (**www.nabcep.org**) offers an independent, voluntary certification program for PV and solar thermal system installers. It requires installers to have at least one year of installation experience and pass a four-hour exam. In addition, to become certified, installers must document systems training and installation; sign a code of ethics; and take continuing education courses to be re-certified every three years. The organization also issues certificates to individuals in entry-level positions who demonstrate a basic knowledge of PV and who are beginning their solar careers.

Some states require solar contractors to be NABCEP certified. In Maine, Minnesota, and Wisconsin, for instance, eligibility for state rebates requires that your PV system be installed by a NABCEP-certified professional. Massachusetts, California, and Delaware rebates "prefer" that a NABCEP-certified professional do the installation of a solar system. Utah requires this certification before issuing a solar contractor license.

Many highly skilled solar energy installers probably do not have NABCEP certification. The Bureau of Labor Statistics does not collect data on solar energy employment, but according to the American Solar Energy Society (ASES), a leading association of solar professionals and advocates, the renewable energy and energy efficiency industries represented more than 9 million jobs and more than $1 trillion in U.S. revenue in 2007. After its September 2009 exam for photovoltaic and solar thermal certification, NABCEP reported that 1,048 individuals now hold the NABCEP installer certification. You can find a complete listing of PV and solar thermal installers who have earned certification on NABCEP's website (**www.nabcep.org**).

TIP: Many solar contractors are relatively new to the business.

The solar power industry has expanded rapidly in the last decade, creating a demand for installers and dealers to sell new solar products. When construction of new homes plummeted after the mortgage crisis of 2008 and jobs became hard to find, many building contractors and tradesmen went into the business of installing solar systems. A number of solar contractors have become licensed or certified only in the last two years. If your solar contractor is new to the business, verify that he or she has the know-how to install your system correctly.

You may live in an area where there are few, if any, licensed or certified solar contractors. If so, a licensed electrician may be able to install your solar PV system. Look for one who has experience with solar systems. In some areas solar contractors are in high demand and you might have to wait several weeks to schedule your installation. If one contractor is readily available when all the others are very busy, there may be a reason: high prices or a reputation for poor work. Be sure to check out the contractor's references.

Comparing prices

A study conducted by the Lawrence Berkeley National Laboratory found that installation costs of PV systems vary widely across the states. Average costs range from a low of $7.60 per watt in Arizona to a high of $10.60 per watt in Maryland. The study suggests that the variation in average installation costs is a result of differences in the size and maturity of PV markets. Greater demand creates more competition, resulting in increased efficiency and possibly reducing the cost of the PV system components. The two largest PV markets in the United States — California and New Jersey — have some of the lowest average costs.

Your cost may be determined by local competition between solar contractors in your area. Tell each contractor that you are shopping around for the best rates. The contractor, who wants your business, might lower his or her price to make the sale.

The national average cost to install a photovoltaic system varies, but can range from $9 to $12 per watt installed, or $9,000 to $12,000 per kilowatt, meaning a 3-kilowatt system could cost up to $36,000. You might expect a smaller system to cost much less, but that is not always the case. Whether small or large, a system entails costs for set-up, permits, and paperwork — as well as equipment and installation — so there might not be much of a price reduction for a smaller system.

Be as specific as possible when asking a contractor to bid on a project. The contractor cannot calculate costs accurately without knowing exactly what components you will need. Most contractors will help you create detailed specifications for your solar systems. Provide one year's utility bills so the contractor can see how much electricity you consume in winter and summer. Make your expectations clear. The contractor should be able to tell you the size and number of solar panels you will need, how much money you will save on your energy bills, and the environmental benefits of your system. Ask for a bid that shows the total installed cost with an itemized breakdown for equipment, labor, permits, taxes, and other costs, as well as deductions for any available rebates or tax credits. Request for a start and finish date for the project, because you will be paying interest on the loan to finance the installation.

When comparing bids, look carefully to see what is included in each bid:

* the quantity, make, and model of the PV panels and equipment
* the system's maximum generating capacity and estimated annual energy production
* warranty information
* what components will be installed where
* hardware
* connection to the grid (if applicable)
* permitting and inspection fees
* sales tax
* travel and transportation
* when the system will be ready for use
* clean-up

Solar water heating system bids should estimate the energy the system will save annually in kilowatt-hours or therms. Solar electric system bids should state system size in watts or kilowatts, and estimate the electricity the system will produce yearly. Ask the contractors to outline the PV system output in AC watts, and indicate the capacity of the system in either watts or kilowatts.

The lowest bid is not necessarily the one you should accept. Consider everything you know about each contractor: years of experience, craftsmanship, customer service, warranty policy, ability to qualify you for rebates and incentives, recommendations from references, and when he or she will be available to do the work. The lowest bid may omit certain expenses or services. Your system could end up costing more money if you accept the lowest bid and then have to pay to fix problems caused by mistakes or poor craftsmanship. A well-established company is more likely to be around several years from now to honor warranties and service your system.

TIP: Installation costs vary from state to state.

If you live in Michigan, do not expect to get the same quote as your cousin in California. Differences in the number of peak sun hours, competition among contractors, the cost of electricity in your area, and the state and local incentives offered to offset the cost of the systems will cause significant variations in the cost of installing a solar system.

Guarantees and Warranties

Some solar rebate programs require that PV systems carry a full two-year warranty from your contractor in addition to manufacturer's warranties on panels and other components. This two-year warranty should cover all parts and labor, as well as the cost of removing and replacing any defective components.

Know your warranties. Many parties are involved in your solar installation: the manufacturers of your solar panels and other components, the dealer who sold the solar panels to the solar contractor, the solar contractor, and if you have back-up batteries, the battery manufacturer. Ask your solar contractor to explain, and perhaps detail in writing, the warranty for each component and who you should contact if there is a problem. Most manufacturers of solar panels guarantee their products for 20 to 25 years. When selecting solar panels, ask your contractor about the manufacturer. The names of some manufacturers are household words, but the solar industry is expanding rapidly and many of the most innovative products are made by unfamiliar companies that have only been in business for a few years. You do not want a 25-year warranty from a company that will disappear in five years.

TIP: Be skeptical of exaggerated promises.

As part of their marketing, some solar contractors offer to guarantee their installations beyond the solar panel manufacturers' warranties — in other words, longer than 25 years. Do not be too impressed. What certainty is there that a company may be around in another thirty years to keep its promise?

Insurance

Installing solar panels on rooftops and wiring electrical circuits can be dangerous work. Ensure that your solar contractor follows standard safety procedures. Ask your contractor's references if they observed any safety violations during their installation. Observe the contractor at work to see if he or she is taking the precautions described below in "Your Safety Comes First."

Many states require licensed contractors to carry certain types of insurance coverage, but that does not mean your contractor's insurance policy has not lapsed. Ask your solar contractor to show you proof that he or she has comprehensive general liability insurance, including public liability coverage and workers' compensation insurance. You do not want to find out after someone has been hurt in an accident that the contractor's insurance will not cover it.

A contractor might be working without the proper licenses. In some states, you can be held liable for an injury of an unlicensed contractor that occurs on your property — even if you were not aware the contractor did not have a license. An unlicensed contractor and his or her employees become your employees, and an injured worker can sue you if he or she is injured on the job. Check with your homeowner's insurance agent about the extent of your liability.

Your homeowner's insurance might cover your injuries if you are installing a system yourself. If you hire workers to help you, and they are not covered under a contractor's policy, purchase worker's compensation insurance to cover them while they are on the job.

Some homeowners insurance will not cover damage caused to the home by a solar water heater or PV system that was not installed by a licensed professional.

A builder's risk insurance policy will cover injuries and damage claims resulting from events that occur during the construction of a new solar home such as fire, heavy winds, or theft. A builder's insurance policy does not protect you if your contractor takes away your supplies or keeps your payment and never finishes the job.

If you are doing your own solar installation, purchase general liability coverage (GLC) to protect you from damage claims due to any eventuality not covered by your homeowner's insurance. For example, homeowner's insurance might not cover injuries to a pedestrian who is walking by your house when a tool falls off the roof while you are working. Under a GLC policy the insurance company has the right to defend you in court against any damage claims that are brought against you. If the court decides you are liable for the damage, the insurance company will pay the claim. Your insurance agent can tell you exactly what kind of GLC you need for your solar project.

Permits and Building Codes

When you begin planning a solar energy project, it is a good idea to find out what kinds of permits are required by your local building department and familiarize yourself with building codes. Look for this information on your city or county website, call the building department, or pay a visit and talk to a live person. Explain what you intend to do and ask what the permitting process is and whether inspections are required. If a permit is necessary, it might take several days or weeks to have your application processed. Knowing about this ahead of time will prevent unexpected delays after you begin your project.

Building codes might specify where you can locate solar arrays — for example, in some areas the array cannot be visible from the street. Building codes also require the structure of your roof to be strong enough to support the extra weight of a solar array or solar water heater. Solar panels must be secured so that they can withstand winds of 125 miles an hour.

A solar contractor is familiar with the process of getting permits and having inspections done in your municipality, and will typically schedule the installation to take place after the permit has been received.

Your homeowners' association (HOA) might also have specific rules regarding the location of solar arrays. If your neighborhood has an HOA, read the covenants, conditions, and restrictions (CC&R) and make sure your design does not violate the rules.

In addition to local requirements, you should look at state and federal requirements for tax credits or financial incentives. Your local municipality might allow you to install the solar panels yourself, but your state may mandate that your PV system be installed by a certified solar contractor to be eligible for state-level incentives.

The National Fire Protection Association (NFPA) (**www.nfpa.org**), established in 1896, provides consensus codes and standards, research, training, and education to reduce the danger of fire and other hazards. It advocates the National Electrical Code® (NEC®), a basic national standard for the safe installation of electrical wiring and equipment. Most state and local regulations incorporate the NEC and inspect electrical installations for compliance with these standards. Some government codes have additional requirements. In 2000, the International Code Council (ICC) published the International Building Code (IBC) to develop codes that would have no regional limitations.

Your Safety Comes First

Whenever you are doing work around your home, the first priority is your safety and well-being and the safety of the people around you. There is no point to installing an energy-efficient solar system if you are not around to enjoy it afterwards!

Most accidents are caused by carelessness, or because someone in a hurry takes a shortcut. The best way to prevent accidents is to exercise a little self-discipline, follow basic safety principles, and take a few extra minutes to keep your work area clean and organized.

When you are working with solar installations, there are four areas in which you might encounter safety hazards:

* Working with electricity.
* Lifting and transporting heavy solar panels and equipment.
* Climbing ladders and crawling around on roofs and in high places.
* Handling lead acid batteries.

If you are not confident that you can do these things safely, hire a professional contractor to do the installation for you.

Working with electricity

Always treat electricity with respect. Remember that your body is a good conductor of electric current. Inadvertently closing a live circuit, or connecting a live power source to the ground, with any part of your body can result in burns, serious injury, and death. An AC voltage of 240 volts is enough to kill, especially if the current passes through the heart. A few milliamps going through the heart, which can occur when a person grabs a cable with each hand, can be fatal. Wet skin has a much lower resistance. If skin is wet or if electricity enters the body through a cut or open wound, a voltage of less than 40 volts can be lethal, making any household electrical device potentially dangerous. Electric shock causes muscle spasms which can result in a fall from a ladder or other injury.

Individual solar panels generate electricity any time they are exposed to sunlight. A 12V solar panel can give off as much as 22V of power when it is not connected to a load. Solar panels connected in series can quickly reach dangerous levels. When not connected, a 48V solar array could generate voltages of 90V, enough to be lethal to an elderly person, child, pet, or someone with a heart condition.

The output from an inverter can be 120V, the same as current from the grid.

Whenever you work with electricity, observe the following precautions:

* **Keep your work area clean and organized.** Do not leave cables or exposed wires lying around. Know where your tools are. Put away equipment when you are finished using it. Do not let flammable materials accumulate in your work area.
* **Do not start work until you know what you are doing.** Read and understand instructions before you start work. If you are unsure about how to do something, consult a professional. When you are working with electricity, a mistake can be particularly disastrous. You could injure yourself, damage expensive and delicate equipment, or create an electrical or fire hazard in your home that could endanger the lives of your family.
* **Shut off all power to a circuit or device before starting work.** Never do any work on a live circuit.
* **Test the circuit.** Even after you have shut off power to a circuit, do not assume that the circuit is not live.

TIP: How to use a circuit tester

Use a circuit tester, a device with two insulated probes, one red and one black, attached to a small lamp, to test the circuit. Carefully grasping the insulated parts of the probes, touch the bare metal end of the black probe to the grounding conductor or the grounded metal box. Hold it there while you touch the bare end of the red probe to the terminal or bare wire that is normally live (typically a black or red wire or a white wire wrapped in black tape). If the circuit is still live, the tester will light up or give a signal. You can test wall outlets by inserting a probe into each slot.

- **Place a warning label on the switch or circuit panel.** Once you have shut off power to a circuit, put a label or note on the switch to let others know not to turn the switch back on.
- **Use an insulated ladder.** Ordinary aluminum ladders can conduct electricity. When working with electricity, use an insulated fiberglass ladder.
- **Avoid working in wet conditions.** Always make sure your hands and clothing are dry when handling electrical equipment. If you must work where there is moisture and water, wear gloves and rubber boots or shoes. Plug tools and equipment into a GFCI (Ground Fault Circuit Interrupter) outlet or extension cord. A GFCI monitors the current in a circuit as it both enters and leaves an electrical device and automatically shuts off the power if it detects a difference in the two amounts.

First Aid for Electric Shock

1. **Do not touch** someone who is in contact with a live power source with your bare hands. A victim who is grasping a power source that is DC will not be able to let go.
2. **Do not get near high-voltage wires** until the power is turned off. Stay at least 20 feet away — farther if wires are jumping and sparking.
3. If possible, **turn off the power source.**
4. Use a **non-conductive material** such as wood, rubber, plastic, or dry cardboard to move the person out of contact with the power source. If the person is not in immediate danger, do not move them any further until help arrives.
5. Call 911.
6. If the person has stopped breathing, begin CPR immediately.
7. If the person is breathing but unconscious, turn on his or her side, with the head tilted slightly back to keep the airway open and prevent the tongue from blocking the airway.
8. Treat for shock and burns.

Any amount of current over 10 milliamps (0.01 amp) can produce painful to severe shock. Currents between 100 and 200 mA (0.1 to 0.2 amps) are lethal. Currents above 200 milliamps (0.2 amps) produce severe burns and unconsciousness but do not usually cause death if the victim receives immediate attention.

Lifting and carrying heavy objects

Solar panels and batteries can be heavy. Lifting a solar array into place will require at least two people. **Do not hesitate to ask for help when you need it.** Avoid injury by following standard safety precautions:

* **Wear protective boots or shoes.**

* **Check that you have enough space** to maneuver, that the footing is good, and that there are no obstacles over which you might trip or stumble.

* **Feet should be shoulder width apart,** with one foot *beside* and the other foot *behind* the object that is to be lifted.

* **Bend the knees;** do not stoop. Keep the back straight, but not vertical.

* **Grip the load with the palms of your hands and your fingers.** Tuck in your chin to make certain your back is straight before starting to lift.

* **Use your body weight to start the load moving, and then lift by pushing up with your legs.** This uses the strongest set of muscles.

* **Keep the arms and elbows close** to your body while lifting.

* **Carry the load close to your body.** Do not twist your body while carrying the load. To change direction, shift your foot position and turn your whole body.

* **Watch where you are going!**

* **Bend the knees to lower the object.** Do not stoop. To deposit the load on a bench or shelf, place it on the edge and push it into position. Make sure your hands and feet are out of the way when setting down the load.

Ladder safety

Accidents can be easily prevented if you take safety precautions seriously and take time to do the job correctly. A fall from a ladder can result in serious injury or death.

* **Clear people out of the area around your work site.** Make sure that no children or pets wander into areas where they could be struck by falling objects. If you are working near a public area such as a sidewalk, place barriers to keep people away while you are working.
* **Make sure your ladder is suited for the job.** Check for cracks or broken joints. If you are carrying heavy objects, make sure your ladder can bear enough weight. A wood or fiberglass ladder should be used for electrical jobs because aluminum conducts electricity.
* **Place ladder on a stable, even surface.** Never place a ladder on top of another object.
* **Place ladder at a 1:4 angle:** The base of the ladder should be one foot away from the wall it is leaning against for every four feet of height.
* **The ladder should extend at least three feet** past the surface you are climbing onto.
* **Always face the ladder** when climbing up and down.
* **Keep both feet on the ladder.** Do not stand and work with one foot on a rung and the other foot on another surface.
* **Secure tall ladders** by tying or fastening the ladder to prevent movement.

* **Never stand on the top** or the paint shelf of a stepladder.
* **Watch your belt buckle,** if you have one. Keep it positioned so it does not catch on a rung.
* **Wear proper shoes and footwear.** Wear shoes with a good grip that will not slip off your feet. Shoes with reinforced toes will protect you from injury if someone accidentally drops a tool or a heavy object from the ladder.

Working with batteries

Lead acid batteries contain sulfuric acid, which is very corrosive and can cause severe chemical burns and eye injuries. Always handle batteries with care.

* **Wear protective clothing.** Wear coveralls, safety glasses or a face shield to protect your eyes, and protective gloves. Steel-toed shoes will protect you from injury if a battery is dropped. Gloves should be sturdy, cover your wrists, allow you to have a firm grip on the battery, and be made of a material that resists sulfuric acid, such as neoprene.
* **Always keep batteries upright.**
* **Do not drop batteries.** Dropping a battery will probably damage it and could crack the casing. If a battery is dropped, immediately place it in a heavy-duty tray or pan to catch any spillage, and check for damage or leaking fluids.
* **Dispose of damaged batteries as hazardous waste.** If you discover that the battery is damaged, double-bag both the battery

and the tray in polythene plastic bags, seal them, and label as hazardous waste.

* **Handle acid spills with care.** Mop up spills immediately with rags and dispose of the rags in sealed polythene bags marked as hazardous waste. If acid gets on your gloves, rinse them with water and replace them so you will not touch any other part of your body with acid. Dispose of gloves as hazardous waste. If acid gets on your clothes, take them off immediately and dispose of them in a sealed polythene bag marked as hazardous waste.

First aid for acid spills.

Always have a first aid kit at your work site. If acid gets into your eyes, rinse them repeatedly with an eyewash and seek emergency medical attention. If acid comes in contact with your skin, wash it off immediately with water and apply an anti-acid wash, cream, or gel to ease the pain. Seek emergency medical attention. If acid splashes into your mouth, wash your mouth out with milk.

* **Do not smoke near batteries.**

* **Avoid short-circuiting a battery.** If you accidentally connect the two terminals of a battery, the resulting surge of energy will generate enough heat in seconds to spark an explosion or a fire or cause serious burns. Remove rings, watches, and metal jewelry when working with batteries. Keep metal tools and cables away where they cannot fall across battery terminals.

Signing the Contract

The contractor you select will prepare a contract for you to sign. The contract is a legal document finalizing the agreement between you, and the clauses in it can decide the outcome of a lawsuit if you and the contractor end up in a dispute. You may be eager to sign the contract and get started on the project, but think about the consequences if the contractor lets you down halfway through the job. Read the contract carefully before you sign, and consult a lawyer if you have doubts or questions.

A contract should cover the following points:

- The names and physical addresses of the contractor and the party (you) responsible for paying for the project.
- A description of the work to be performed. Typically the detailed plans, including the make and model of the individual components to be purchased, will be attached to the contract.
- Start date and end date for the project.
- Total amount to be paid and payment arrangements.
- Conditions under which final payment will be made.
- Specific duties of the contractor, such as applying for permits and transporting materials to the work site.
- A guarantee that the contractor has the necessary licenses and certifications and is insured and bonded.
- A clause specifying who is responsible for additional insurance and who will receive payment in case of a claim.
- A clause specifying how changes will be made and approved, and who will pay for what.
- Conditions under which either party can terminate the agreement.
- A clause specifying what will happen if the contractor fails to finish the project.
- A clause specifying what will happen if you fail to pay the contractor.

- Arrangements for communication between you and the contractor.
- A clause specifying what will be done if the contractor damages your home or any of the components, or if the contractor makes a mistake installing your system.
- The procedure that will be followed if there is a delay or a change to the plans — for example, if a component is back-ordered or the installation does not pass inspection.
- What will happen if either you or the contractor dies before the work is completed.
- Any other agreement between you and the contractor.

Working with a Contractor

From the beginning of the project, communicate often with your contractor and ask questions if you do not understand what is going on. You and your contractor have a similar goal: a successful solar installation. The contractor wants to finish the job, get paid, and walk away with a good reference. You want a well-crafted solar system that will work efficiently for the next 25 years. Give your contractor a phone number where you can be reached or receive a message during the day if a problem arises. Treat your contractor with respect and courtesy.

Some changes are inevitable during a solar project as structural problems or errors in the original design are discovered, components are unavailable, or newer, less expensive models come on the market. You and your contractor need to agree on a procedure for approving and making changes. Ideally changes should be made while the project is still in the design stage and the only cost is additional time to research and re-draw the plans. Taking a little extra time to review the design at this stage saves you money in the long run. Relocating solar panels during installation will require extra work. Changes to one part of the system may affect other components or other parts of the building if you are constructing a new home. For example, if you decide to add more solar panels, you may need to buy a different inverter. Making changes after an installation is complete probably involves discarding materials and repeating work.

TIP: Avoid becoming emotional when communicating with your contractor.

You will not accomplish what you want by becoming angry and hostile when problems or delays arise. Instead, your contractor will dread telling you when something needs to be changed or done differently, and may avoid telling you the truth. Remain calm and reasonable and adopt a problem-solving attitude.

Do not make a final payment to the contractor until after the job is complete and has been inspected and approved. It may be necessary to release partial payments as the job progresses if the project takes more than a few days, or to pay for purchasing solar panels, system components, and materials. In that case, payment terms should be specified in your contract and payments to the contractor should be contingent on the accomplishment of certain milestones. The lender financing your project should be informed and the arrangements should be included in your loan contract.

Designing Your Solar Installation

Before you can make a detailed design of your solar project and estimate the total cost, you must gather several important pieces of information. Whether you are creating a system to run a single appliance or supplying all the electricity for your household, you need to estimate as accurately as possible how much power is required, and whether that power should be AC or DC. You also need to determine how much energy you can expect your system to produce in your climate and your particular physical circumstances. You must look at your home and decide where you will locate the various components of your system. The final step is a shopping list of the components and equipment to create a system that supplies the energy you need at a reasonable price. Then you will be ready to plan the installation, including applying for permits, arranging for the delivery of materials, and finding people to assist you with the heavy work.

If you are hiring a solar contractor, he or she will probably do most of this work for you before submitting a quote for the job. Your contractor is already familiar with local conditions, building codes, the different solar products available, and the hardware and equipment needed for installation. It is a good idea for you to have an understanding of what the solar contractor is doing even if you do not go into all the details. You will know what he or she is talking about and you will be better able to compare bids and select the best contractor for the job.

If you are planning a do-it-yourself project, you are about to embark on a long and interesting learning process. This book is only a stepping-off point. The following chapters will explain the basics. If you have never done this kind of work before, you will probably find yourself consulting technical manuals, instruction booklets, the Internet, your manufacturer's customer service representatives, local home improvement stores, and your friends and neighbors for clarification and advice. Fortunately a vast amount of information is available online. *See Appendix 3: Useful Websites.*

Doing a Load Analysis

The first step in the design of a photovoltaic system is finding out exactly how much power you need. For a PV system that will power only one or two appliances, such as a well pump and a light, you need to know how much voltage those appliances require. For a standard grid-tie PV system that will connect to your utility system, you need only a 12-month energy usage report, like the one described in Chapter 3: Understanding your electricity bill. Look at a year's worth of electricity bills or request an energy usage report from your utility (if you view your electricity bill online, it may be available on the utility company website). Your monthly bill may even include a figure for your average daily usage based on the past 12 months.

The table below, from EcoSolar, a solar contractor in Oregon, shows the optimal size for a grid-tie PV system in Klamath Falls, Ore. The calculation is based on the assumption that Klamath Falls receive an annual average of five hours of sunlight per day, and your system is located where its efficiency is not diminished by any shade during the day.

Optimal PV system size according to meter usage for Klamath Falls, Ore.

Find this on your power bill -- kWh / Day Usage (Annual Avg.)	Optimal PV System size (Watts)
9.4	2,000
14.1	3,000
18.8	4,000
23.5	5,000
28.2	6,000
32.9	7,000
37.6	8,000
42.3	9,000
47	10,000
70.5	15,000
94	20,000
188	40,000
*These numbers are based on 5 peak sun hours per day on average annually and assume a 94 percent total system efficiency with no shading.	

The optimal size for your grid-tie system will depend on the average hours of sunlight per day in your geographical location. If you are designing an off-grid PV system or a PV system for a new home, you will need to do a complete load analysis for your home. This involves making an itemized list of each appliance that uses electricity in your home, including digital clocks, cell phone and camera chargers, and smoke alarms. You then calculate the amount of power needed to operate each one. Use the worksheet below to do a load analysis. It will take some time but is easy to do.

Load analysis worksheet

1. **Appliance.** Using a spreadsheet or lined paper, make a chart like the one shown here. You can download an Excel spreadsheet with the formulas for calculations already included from Home Power Magazine's website (**http://homepower.com/webextras**). Go through your house, room by room, listing each appliance that uses electricity. Do not forget the power tools in the garage or electric tools that you use for your hobbies.

2. **Quantity.** How many of each appliance do you operate? For light bulbs, assume that each member of the family turns lights on and off when he or she enters and leaves a room. Count one light bulb for each person in the household, and estimate how many hours those lights are in use. Lights that are left on all the time or for extended periods should be counted separately. Lights of different wattages should also be separate entries.

3. **Volts.** Most AC household appliances run on 110V, but a few such as power tools and clothes dryers run on 220V. Battery-powered electronics like laptop computers, cameras, and shavers run on 12V, 24V, or 48V and typically come with a charger or transformer that converts AC current to the right DC voltage. If you are building a new house or buying new appliances, you might find a refrigerator or TV that runs on 12V DC.

You can find the voltage of your appliances in several ways:

* Read the label on the back of the appliance or on the power supply. The voltage may be listed as 120 volts, 120V, 120 volts AC, or 120 VAC. This is the maximum voltage at which the appliance runs.
* Check the manual.
* Consult a list of typical power consumption for similar appliances. (See chart below.)
* Use a wattmeter to measure the watts actually being used as the appliance operates. This is the most accurate measure, because many appliances have a range of power settings (such as volume settings or heat settings on a stove) or have motors that switch on and off.

About Wattmeters

A wattmeter measures the instantaneous power (watts) and the total energy used (watt-hours or kilowatt-hours) by an appliance. It plugs into an electrical outlet and then the appliance plugs into the meter. A digital electronic wattmeter meter samples the instantaneous voltage and current thousands of times a second and averages the amounts to give you a reading. The simple models show the reading on an LCD display. Wattmeters are especially useful for giving an accurate reading on appliances that run in cycles, like dishwashers and washing machines. A simple wattmeter costs $25 (Kill A Watt) to $100 (Watts Up Meters — **www.wattsupmeters.com**) and can be purchased online or at a home improvement store.

Wattmeter: A simple Kill A Watt wattmeter will give you an accurate measurement of the power used by your appliances.

4. **AC or DC.** Note whether the appliance runs on AC current from a utility or DC power from a battery or transformer. This information will be useful later if you are designing an off-grid system.

5. **Inverter Priority (IP).** Appliances that are automatically turned on and off by thermostats or timers, like refrigerators and air conditioners, take priority over appliances that you can turn on and off at will. This is because you cannot control when they might draw electricity from your system. Appliances like washing machines and hair dryers do not have priority because you can deliberately avoid using them at times when your system is already at capacity.

6. **Run Watts.** The watt rating shown on an appliance label typically represents the greatest amount of power used by the appliance when it is running at its top setting. You can safely reduce this number by 25 percent for appliances that you never use at their highest settings, such as your television or speaker system. Some appliance labels list only the current (amperage) and voltage. Amperage may be expressed as 0.5 amps, 0.5 A, or 500 mA. To calculate run wattage, simply multiply the volts and the amps.

$$\mathbf{W\ (power) = V \times A}$$

You can obtain the most accurate estimates by using a voltmeter. When measuring the current drawn by an appliance using a motor, the meter will show about three times more current in the first second as the motor starts up than when the motor is running.

7. **Hours per Day.** Write down how many hours a day the appliance is turned on, on the days when you use it. Appliances controlled

by thermostats or timers have "duty cycles" when they are running and other times when they shut off. You will either need to observe when they are turning on and off during the day, or use a voltmeter to find how much power they use. Remember that these appliances may run for longer hours in summer or winter.

8. **Days per Week.** Write down how many days per week the appliance is in use.

9. **Phantom Load.** Appliances that consume power when they are turned off are called "phantom loads." Note whether the appliance has a phantom load.

TIP: Do not overlook phantom loads

Phantom loads use small amounts of electricity 24 hours a day, and over time these small amounts add up.

Examples of phantom loads are appliances with digital clocks or timers, appliances with remote controls, chargers, and appliances with wall cubes (the little boxes with power cords that plug into AC outlets). Wall cubes consume 20 to 50 percent of the appliance's total wattage when the appliance is turned off. To control phantom loads, unplug these appliances when not in use, or plug them into a power strip that can be turned off with a master switch.

10. **Watt Hours per Day.** Use the information from previous columns to calculate how much electricity each appliance uses on an average day.

Average Watt-hours per Day =
Quantity × Run Watts × Hours per Day
× Days per Week ÷ 7 Days

Add up all these amounts. The total is the **daily rate of electricity consumption** that your off-grid PV system will have to support, which will determine the cost of your system.

TIP: Do not be discouraged by high electricity consumption.

You may be alarmed when you see how much electricity is consumed in your household, and the size and expense of the system needed to provide it. Do not give up yet! You can do many things to reduce the load on your PV system, including purchasing more energy-efficient appliances, replacing an electric stove with a propane stove, and not running all your appliances at once.

11. **Percent of Total.** Using your figure for total daily consumption of electricity, calculate what percentage of the total is represented by each appliance.

Percentage of Average Daily Load =
Individual Average Watt-hours per Day ÷
Total Average Watt-hours per Day

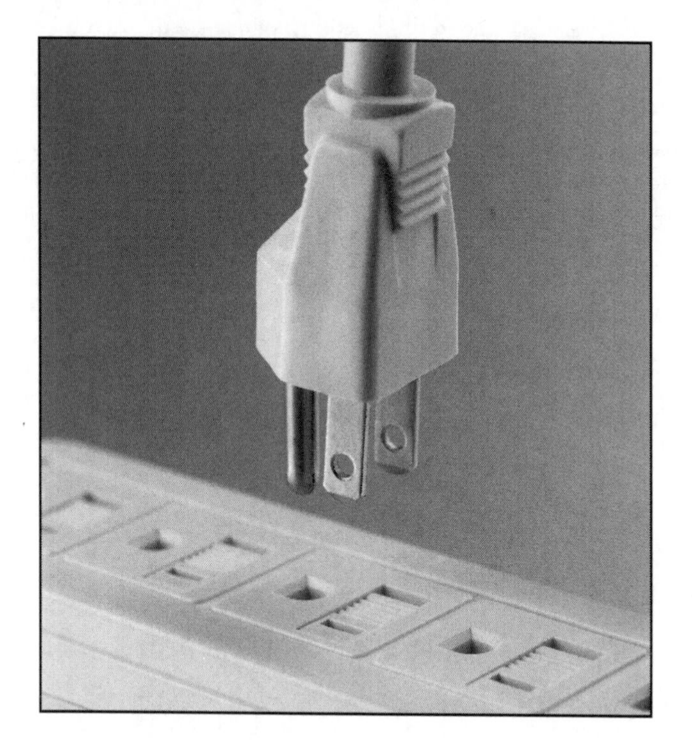

Sample Load Analysis Worksheet

Appliance	Quantity	Volts	AC or DC	IP Y/N	Run Watts	Hours/ Day	Days/ Week	Ph-L Y/N	W-hours per day	Percent of Total
Refrigerator (sample)										
TOTAL										

Average Wattage of Common Appliances

Appliance	Watts
Aquarium	50–1210
Clock radio	10
Coffee maker	900–1200
Clothes washer	350–500
Clothes dryer	1800–5000
Dishwasher (using the drying feature greatly increases energy consumption)	1200–2400
Dehumidifier	785
Electric blanket-Single/Double	60 / 100
Fans	
Ceiling	65–175
Window	55–250
Furnace	750
Whole house	240–750
Gas Dryer	300–400
Hair dryer	1200–1875
Heater (portable)	750–1500
Clothes iron	1000–1800
Microwave oven	750–1600
Personal computer	
CPU - awake/asleep	120 / 30 or less
Monitor - awake/asleep	150 / 30 or less

Appliance	Watts
Laptop	50
Printer - Ink Jet	50
Printer - Laser	600
Radio (stereo)	70–400
Refrigerator (frost-free, 16 cubic feet)	725
Televisions (color)	
19″	65–110
27″	113
36″	133
53″-61″ Projection	170
Flat screen	120
Toaster	800–1400
Toaster oven	1225
VCR/DVD	17–21 / 20–25
Vacuum cleaner	700–1440
Washing machine	200–900
Water heater (40 gallon)	4500–5500
Water pump 1/2 HP	500–900
Water pump - 1 HP	900–1500
Water pump (deep well)	250–1100
Water bed (with heater, no cover)	120–380

System inefficiencies

In addition to the electricity drawn from your system by your appliances, a certain amount of power is lost along the way due to inefficiencies in the batteries and inverter and resistance in the circuits. To ensure that your solar array will produce enough electricity for your needs, you will have to allow for this additional loss.

Batteries do not discharge 100 percent of the energy they store when they are charged. Charge Cycle Efficiency is a measure of the percentage of energy you can draw from the battery compared to the amount of energy used to charge it. Your battery manufacturer can supply you with the information, but the Charge Cycle Efficiency for standard industrial batteries is typically 95 percent. Whether you are designing an off-grid system or a grid-tie system with battery backup, the energy from your solar panels is stored in batteries and then fed to your electrical circuits. You will need to compensate by adding another 5 percent of your total watt-hours per day to your load analysis.

If your system is using an inverter to convert DC to AC, it typically loses about 10 percent of the energy flowing into it. You can get exact specifications from the manufacturer, but you would be safe adding another 10 percent of your total watt-hours per day to your load analysis.

Low-voltage energy is also lost when it travels through cables for a distance. If you are installing a solar array on the ground or on a separate shed, you will also need to allow for loss due to resistance.

Completing a load analysis has given you the first spec (specification) for your PV system design — exactly how much energy it needs to produce. This is the final goal of your PV system design. The load analysis also shows which appliances use the most electricity and where you might be able to make changes that will improve the energy efficiency of your home.

How Much Solar Energy is Available?

The output listed on the label of a solar panel is the amount of energy produced when that panel is exposed to full sunlight at solar noon in the middle of summer on a clear day. Depending on where you live, those conditions exist only for a short period of time, and not at all on cloudy and stormy days. The total amount of energy a solar panel can produce depends on the irradiance of its location. *The solar radiation maps in Chapter 3 show the average winter and summer irradiance for regions of the United States.* You need to know what the irradiance is for your specific location. NASA has been collecting data on solar irradiance for 22 years, and this data can be used to calculate the average monthly irradiation for your neighborhood. The National Renewable Energy Laboratory (NREL) PVWatts™ calculator (**http://pvwatts.nrel.gov**) determines the energy production and cost savings of grid-connected photovoltaic energy systems throughout the world. Solar Panels Plus, LLC (SPP), a manufacturer, importer, and distributor of solar power products, has a Zip-Code Solar Insolation Calculator (**www.solarpanelsplus.com/industry-professionals/insolation-charts**) that will give you the average monthly solar irradiance for your zip code. The Solar Electricity Handbook website (**http://solarelectricityhandbook.com/solar-irradiance.html**) has compiled solar irradiance data for cities and towns all over the world. Look up the solar irradiance in your area for each month of the year. Multiply the solar irradiance times the output wattage of a solar panel to calculate how much energy it can be expected to produce each month.

Below is the average monthly irradiance for Orlando, Fla. and Salt Lake City. You can see that though the yearly average irradiance is almost the same for both locations, there is a considerable difference between the estimated output in Florida and the estimated output in Utah during winter and summer months.

Monthly Solar Irradiance and Estimated Output in Wh for Florida and Utah

	Orlando, Fla. Latitude: 28.4, Longitude: -81.4		Salt Lake City Latitude: 40.7, Longitude: -111	
MONTH	Irradiance in kWh/m^2*day	Estimated Output in Wh	Irradiance in kWh/m^2*day	Estimated Output in Wh
January	3.29	65.8	2.36	47.2
February	3.93	78.6	3.2	64
March	4.92	98.4	4.45	89
April	5.9	118	5.48	109.6
May	6.27	125.4	6.44	128.8
June	5.56	111.2	7.22	144.4
July	5.46	109.2	7.01	140.2
August	5.09	101.8	5.98	119.6
September	4.59	91.8	5.11	102.2
October	4.19	83.8	3.77	75.4
November	3.57	71.4	2.52	50.4
December	3.08	61.6	2.13	42.6
Yearly Average	**4.65**	**93**	**4.64**	**92.8**

To estimate the size of the solar system you need to meet your energy requirements, take the total watt-hours per day from Column 10 of your Load Analysis Worksheet and divide it by the lowest average irradiance for the year. This will give you the approximate size of the PV system you need. If the house will not be occupied during the winter months you can exclude them and choose the month with the lowest irradiance during the period when the house will need electricity.

Total Watt-hours per Day ÷ Lowest Average Irradiance = Size of PV System

For example, if you live near Orlando and you have calculated that your home uses an average of 800 kWh per day:

800 ÷ 3.08 = 259.74

You need a 260-watt solar array.

System inefficiencies

A 260-watt solar array, however, may not produce 260 watts all the time. Discrepancies between the voltage of the solar array and the voltage of the batteries being charged by the electric current can prevent the solar array from producing its peak output.

Power point efficiency is a measure of the ability of your solar array to produce its optimum power output. The rating on a solar panel is its "peak power output," or "peak power voltage" (Vpp), the voltage at which a 12-volt solar array, or PV module, can produce its maximum output in bright sunlight, typically 15 to 22 volts. Vpp varies with the amount of sunlight falling on the array and the temperature of the array. For a PV module to charge a battery, its voltage must be slightly higher than the voltage of the 12-V battery. If its voltage is lower, the output from the PV module will drop. To compensate for this, typical solar panels are designed to produce a Vpp of 17V at a temperature of 25°C (77°F). At higher temperatures, the solar panels may produce only 15V, still higher than a 12-V battery. At cooler temperatures, the solar panel becomes more efficient. On a cool, sunny day the solar panel could produce an output of 18V. If the voltage from the PV module is too much higher than the battery, it will be dragged down to the capacity of the battery and the extra potential to produce energy will be lost.

The voltage of the batteries in a 12-volt system can range between 11V and 14.5V, depending on the state of charge of each battery and whether appliances are drawing current from the system. On a cool day when your solar array is capable of producing 18V but the voltage in your batteries is hovering around 12V, that potential to produce an additional 6V of energy could be lost.

Maximum power point tracking (MPPT) is a technology that tracks the fluctuations in the Vpp of the solar array and varies the ratio between the voltage and current delivered to the battery, to deliver maximum power. Excess voltage from the solar array is converted to additional current to the battery.

MPPTs are incorporated in both inverters and controllers, and even in some solar panels. Inverters and controllers with MPPT are more expensive and are not always cost-effective. An MPPT controller pays for itself if your system produces more than 120 watts of power. An MPPT inverter is only needed in grid-tie systems where the solar panel feeds current directly into the inverter. An MPPT inverter becomes cost-effective when your PV system produces more than 300 watts of power.

Even with MPPT, some of the power coming from your solar panels is inevitably lost in transition due to power point inefficiencies. You must increase the output of your solar array to compensate for these losses. If you are using MPPT, divide your required output by 0.75. If you are not using MPPT, divide it by 0.9.

Using the example from above, the 260-watt solar array:

MPPT controller	**260 ÷ 0.9 = 288.89 = 289 watts**
Non-MPPT controller	**260 ÷ 0.75 = 346.67 = 347 watts**

To supply 260 watts of electricity, you will need a 347-watt solar array if you do not have MPPT.

In addition, all solar panels lose approximately .05 percent of their efficiency each year as they age, or 10 percent during the 20-year warranty period. Therefore it is better to slightly overestimate the size of your solar array.

Compensating for high temperatures

Solar panels generate power less efficiently as they heat up. When solar panels are given a rating, they are tested at 77°F (25°C) with an irradiance of 1,000 watts per square meter. A 20-volt solar panel will produce more than 20 volts at temperatures below 77°F (25°C). As the temperature of the solar panel rises higher and higher above 77°F (25°C), the efficiency of the solar panel decreases. Solar panels will produce more power on a cool, windy day than in the blazing sunlight of a hot summer day. The dark faces of solar panels absorb quite a bit of heat, and rooftop solar panels can reach temperatures of 160 to 170°F (80 to 90°C) during the summer in a hot region.

Your solar panel manufacturer will typically provide a "temperature coefficient of power" rating — the percentage of total power reduction for each 1°C increase in the temperature of the solar panel. This is usually around 0.5 percent, meaning that for every degree that the temperature of the solar panel rises, its output is reduced by .05 percent.

For a PV system in a cool climate, or an off-grid system that generates a lot of excess electricity during hot summer months, you do not need to compensate for inefficiency due to heat. If you live in a region where the temperature is above 77°F (25°C) for much of the year, you will need to factor this inefficiency into your calculations.

The table below shows how temperature affects the performance of a solar panel:

TEMPERATURE	5°C/ 41°F	15°C/ 59°F	25°C/ 77°F	35°C/ 95°F	45°C/ 113°F	55°C/ 131°F	65°C/ 149°F	75°C/ 167°F	85°C/ 185°F
Output	110w	105w	100w	95w	90w	85w	80w	75w	70w
Percentage loss	10%	5%	0%	-5%	-10%	-15%	-20%	-25%	-30%

Effect of temperature on the efficiency of a 100-watt solar panel (assuming a temperature coefficient of power of .05 percent)

Look up the average summertime temperatures for your region. If yours is a rooftop installation and the air temperature is 25°C (77°F) or more, multiply the air temperature by 1.5 to get an estimate of the temperature of the solar panels. For pole-mounted installations, multiply the air temperature by 1.2. Then increase the wattage requirements of your system by the percentage shown in the table to compensate for inefficiency caused by heat. If the temperature coefficient of power provided by your solar panel manufacturer is lower than .05 percent, adjust the percentage accordingly.

You can keep your rooftop solar panels cooler by mounting them with a gap of 3 to 4 inches between the panel and the roof to allow air to circulate underneath. Mounting your solar array on a pole allows air to circulate around them and cool them slightly. Some solar panels are also manufactured with built-in cooling systems that circulate cool air or liquid underneath.

Estimating the Cost of Your System

Here is a review of how you calculate the size of the PV system you need for an off-grid system:

1. Do a load analysis and find the total number of watt-hours per day that your system must produce to meet all your energy needs.

2. Look up the solar irradiance for your geographic location and find the lowest average irradiance for the months when you will be using your PV system. Divide the total watt-hours per day by the lowest average irradiance to get the size of the PV system that will output enough watts to supply your energy needs during the time period with the least amount of available sunlight.

3. Increase the size of your PV system to compensate for power point inefficiencies.

4. Increase the size of your system to compensate for inefficiency due to high temperatures.

Now that you know the size of the system you need, you can roughly estimate how much it will cost. In 2010, the average cost across the United States for a grid-tie system was $8 to $9 per watt for a system installed by a professional contractor. An off-grid system or one with back-up batteries costs 20 to 30 percent more, or about $10. If you installed the system yourself, the cost was around $6 to $8 per watt.

Grid-tie systems tend to cost a little more because you need a sophisticated inverter and a licensed electrician to install it. Smaller systems cost more per watt than large systems because both systems require the same basic equipment and have the same fixed costs.

Do not be alarmed if the initial estimate seems very high. In many areas, rebates, incentives, and tax deductions will cut 30 to 50 percent off the cost of your system. *You can also implement some of the measures discussed in Chapter 4: Making Your Home Energy Efficient to reduce your consumption of electricity.*

TIP: Owners of off-grid systems are extremely conscious of their energy use.

Homeowners who rely entirely on an off-grid PV system for all their electricity watch their electricity consumption carefully and become very conscious of the energy consumed by each appliance. They know that excessive or unnecessary use of energy could tax their system beyond its capacity.

Selecting a Site

With an accurate estimate of the amount of power your system must produce, you can now calculate how much space you will need for your solar array and find the best location for it on your property.

An amorphous solar panel one meter square (about 9.9 square feet) generates about 60 watts. A crystalline solar panel the same size generates about 160 watts. Depending on which technology you plan to use, work out how many square feet of space you will need for your array by dividing the total size of the system by 60 or 160. Multiply the result by 9.9.

Doing a site evaluation

The purpose of a site evaluation is to find the best possible location for your solar array and the other components of your system, and to identify any trees, buildings, or other obstacles that might shade that location during some part of the year. A solar contractor does a site analysis using sophisticated tools such as the Solar Pathfinder and accompanying software ($400 to $500) (**www.solarpathfinder.com**); or the $2,000 SunEye 210 by Solmetric (**www.solmetric.com/buy210.html**), a handheld device that incorporates a digital camera with a fisheye lens and a computer to capture and analyze information about a solar site.

Because your site evaluation is a one-time project, you will probably prefer to do a manual site evaluation. To do a preliminary site evaluation you will need paper and pencil, a protractor, a carpenter's level, a compass, a tape measure, and a ladder to climb onto the roof.

TIP: Tape pieces of cardboard together to make a panel the size of your solar array.

It is helpful to make a cardboard panel the size of your proposed solar array so you can see its actual size and try positioning it where you want to locate the array. This will alert you to potential problems and adjustments that need to be made.

1. **Confirm that your property has adequate access to sunlight.** If you are in the Northern Hemisphere, stand at the east side of the property looking south and to the west for major obstacles. If you are in the Southern Hemisphere, stand on the west side and look north and west. If you are close to the equator the sun will pass directly overhead and only obstacles to the east or west will be a problem. Identify possible locations for a solar array.

2. **Draw a rough sketch** of your house and any trees, sheds, or light poles, including trees and obstacles on adjacent properties.

3. **Locate the best site for your solar array.** The array will need to be mounted at an angle facing into the sun. This angle varies throughout the year, but in the US it is between 40° and 64°; in Canada between 20° and 40°, and in the UK between 33° and 40° from vertical. Determine how you will mount your solar array.

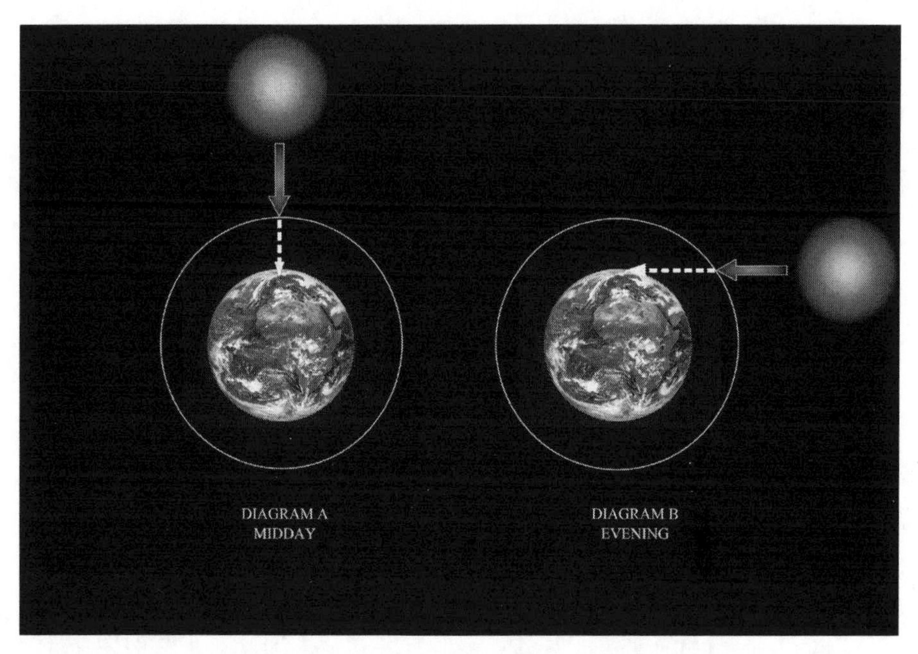

The angle at which the sun's rays strike your solar panels changes throughout the day.

Roof mounting:

A flat roof or a south-facing roof is an optimal location for a solar array. If the roof does not slope directly to the south, you can use angled supports to face the array directly into the sunlight.

Roof panels have to be installed where there are no shadows during the day. Because this house has no south-sloping roof, this 2.5-kilowatt system is installed on an east-sloping roof with an additional tilt toward the south.

Measure the slope of your roof in degrees by holding a protractor against the angle where a rafter under the roof joins one of the cross beams. You can also use an expensive tool called a "roof angle finder" available in home improvement stores. Press the angle finder against one of the roof rafters and it will give you a reading in degrees.

Check under your roof to see if it is strong enough to support a solar array. Individual solar panels weigh about 30 to 40 pounds, and the weight can add up. Determine what kind of hardware and mounting fixtures will be needed to attach the array to the roof. Consult an architect or a builder if you are uncertain. Roof mounting kits are sold by solar panel manufacturers, and you can also build your own.

Solar arrays need to be cleaned off occasionally because dust, grime, and snow all block sunlight and make the solar panels inefficient. Consider how you will clean solar panels mounted on your roof without too much difficulty. Mounting them on the lower edge of the roof makes them easier to reach with a ladder instead of having to climb around your roof.

Measure the available space where your array will be situated and sketch it in on your drawing. Put your cardboard cutout in place to identify any potential problems with installation. Look at roof vents and gutters and determine where runoff from the solar panels will go when it rains or snows.

Wall mounting:

Solar panels can also be mounted on a wall if they can be angled in the right direction without protruding too far from the wall. Make sure no overhanging eaves or architectural features will interfere with sunlight falling on the panels.

Wall-mounted solar panel.

Ground mounting:

A solar array can be mounted on a frame on the ground, where it is easy to clean and you can manually change the angle of the array throughout the year. A ground mount requires adequate open space in your yard or garden. Take note of any obstacles that might cast shade on your installation. Solar suppliers sell a variety of ground frames or you can build your own. Depending on ground conditions, you might have to lay a foundation to support the frame.

Ground-mounted solar array at a home in Colorado.

Mounting on a shed or patio:

A solar array can be located atop a specially constructed storage shed or patio awning. PV panels incorporated in building materials and components can take the place of traditional panels. You will need to calculate the cost of building the shed or patio in addition to your solar installation.

Pole mounting:

Pole mount

Smaller arrays of up to 600 watts can be mounted atop a pole. You will need a heavyweight pole and a substantial foundation to withstand heavy winds. Pole mounts help to keep the solar array cool in a hot climate. Solar suppliers sell poles and hardware.

Solar trackers:

A solar tracker is a mechanical device that changes the angle of the solar panels to follow the path of the sun, so your solar array is always at the optimum angle. It can increase the capacity of your system by 30 percent. Commercial solar trackers are expensive, so it is usually less expensive to buy a slightly larger array that will generate more power.

If you enjoy do-it-yourself projects, you can build your own solar tracker. You will find instructions for building your own solar tracker on websites like these:

* DIYDaSolar.com Network (**http://diysolar.dasolar.com**)
* LivingonSolar.com (**www.livingonsolar.com/solar-tracking.html**)
* GreenWatts (**http://solartracker.greenwatts.info**)

Tilt and Orientation Factor (TOF)

Tilt and orientation factor (TOF) measures how well the position of a solar array maximizes its production capacity by combining the angles of panel tilt and orientation. A TOF of 100 percent indicates that a solar array's position fully maximizes its power production capacity. A TOF of 70 percent indicates that the system is placed in such a way that it loses 30 percent of its production capacity.

An online Compound Angle tool by Solmetric (**www.solmetric.com/compound-angle.html**) can be used to work out the exact angle and tilt for your solar panel array.

TIP: Horizontally mounted panels should never be tilted less than 5°

Horizontal panels should always be tilted slightly so rain or a hose can easily rinse off accumulated dirt.

4. Look for obstructions that may cause shading

Now that you have decided on a location for your solar array, you must look for any obstacle that will obstruct the sunlight falling on it during any time of the year. Shade is the nemesis of a PV system. Solar panels are constructed in such a way that one under-performing cell diminishes the output of the entire panel. The most efficient solar panels in the world will operate at only a fraction of their capacity if they are partially shaded.

To illustrate the effect of shade on a solar panel, a business card covering half of one solar cell can reduce the panel's production by 27 percent. The shadow cast by a chimney, vent pipe, utility pole, tree, or building can severely hamper the performance of your solar array. It is important to identify these obstacles now during the design stage, because they will prevent your solar system from operating as expected.

Shading that occurs before 9 a.m. or after 3 p.m. can cause your system to lose about 20 percent of its capability to produce power in summer, and 10 percent in winter months. If shading occurs both before 9 a.m. and after 3 p.m., you could lose 40 percent of your capacity. Persistent shading of even one or two cells in a solar panel can damage your system by creating hot-spot heating.

Shading during peak production hours between 9 a.m. and 3 p.m. will severely limit your system's capacity. You should seriously reconsider your solar project if you cannot avoid shade during these hours.

Tracking the path of the sun

The sun's path through the sky varies throughout the year, as the Earth tilts on its axis. When determining whether an obstruction will shade your solar array, you must consider the sun's path during every season of the year.

The position of the sun is plotted using two measurements — azimuth and elevation. Azimuth is a horizontal coordinate for locating the sun in the sky, expressed as the number of degrees clockwise from true north. Azimuth can be determined using a compass. Elevation is a vertical measurement of the sun's position in the sky, measured in degrees as angle between the horizon and a line drawn through the sun.

Around March 21 and September 22 of every year, the sun rises due east of the equator and sets due west of the equator. These days are called "solar equinoxes." At solar noon on the equinox, the angle of the sun is 90° minus the local latitude. At the winter solstice, which falls on December 21, in the Northern Hemisphere, the angle of the sun is 23.5° lower than at the equinox. At the summer solstice, which falls on June 21, the angle of the sun is 23.5° higher than at the equinox. The online Solar Calculator (**www.esrl.noaa.gov/gmd/grad/solcalc**) of the National Oceanic Atmospheric Administration's (NOAA) Earth System Research Lab (ESRL) gives the angle of the sun's elevation for any location on earth on any day of the year.

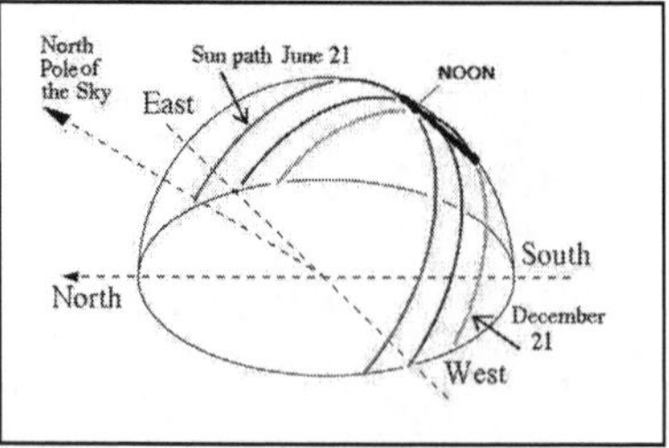

Path of the sun from the Northern Hemisphere at different times of the year

The illustration to the right shows the path of the sun during the equinox and winter and summer solstices, as seen from the Northern Hemisphere. Imagine that you are standing, facing south, on the site of your solar array at the intersection of the lines that cross East to West and North to South.

To verify that no obstacles will obstruct the sunlight falling on your solar array at any time of the year, use an ordinary protractor. Lay a pencil pointing from the center of the protractor along the angle of the sun in winter. Using a compass, find due south. Standing on the intended location of your solar array, look south along the pencil and turn from east to west, checking for any obstructions. If you find an obstruction such as a utility pole, tree, or building that will cast shade on your array, you will have to either relocate it or find a way around the obstacle, perhaps by elevating your array on a pole or frame.

TIP: Use a sun chart to do a site survey

If you are concerned that surrounding trees and obstructions may create problems for your solar array, you can do a more detailed survey using a sun chart — a graph of the sun's elevation and azimuth throughout the day. The Solar Radiation Monitoring Laboratory at the University of Oregon offers a free program that creates printable sun charts for any location on its website, **http://solardat.uoregon.edu/SunChartProgram.html**, and instructions for using a sun chart to do a site survey. You will need a compass and a clinometer — an instrument that measures elevations (your protractor and pencil are a simple clinometer). You will be able to plot obstacles on the sun chart and see whether they present a serious difficulty.

TSRF

Total solar resource fraction (TSRF) measures the efficiency of a solar array by combining the TOF with the percentage of shading on it during the year. The percentage of shading can be calculated using a sun chart (see below) with a percentage assigned to each section of the chart. The closer the TSRF value is to 1, the more efficiently the array will produce solar electricity.

TIP: The government wants your system to be efficient.

To receive many state and local rebates, a solar system must meet a certain TSRF standard. The government does not want to help pay for an inefficient solar system. For example, Oregon Administrative Rule #330-135-0030, "1.5 Percent for Solar Energy in Public Building Construction Contracts," mandates a TSRF value of 0.75 or higher.

Trees and landscaping

The length of a shadow is determined by the elevation of the sun, and the direction of the shadow is determined by the azimuth. The higher the elevation of the sun, the shorter the shadow. A tree's shadow will always be opposite the sun, and longest around sunrise and sunset. During the day, the tree casts its shortest shadow at solar noon. In the Northern Hemisphere, a tree casts its longest shadow on December 21, the winter solstice. The shadow then gradually shortens until June 21.

If you determine that a tree is going to significantly shade your solar array during part of the day, you may have to make a choice: relocate your solar array or cut down the tree.

Trees shading a solar array.

Trees near your solar array will have to be regularly maintained to prevent damage from falling branches and the obstruction of your solar panels by drifting tree pollen, flowers, and fallen leaves.

If you are building a new home, plan your landscape carefully. For the south side of your home, select trees and shrubs that will not grow tall enough to obstruct your solar array.

Is access to sunlight a legal right?

As increasing numbers of homeowners install solar arrays on suburban roofs, the question of legal protection for solar access must inevitably arise. It takes a homeowner as long as 20 years to recoup the cost of a solar installation. To operate effectively, solar panels must have unrestricted access to sunlight, but the path of the sun's rays often passes through a neighbor's airspace. Many homeowners are reluctant to invest in a solar system because they fear their neighbors might plant trees or construct buildings that would obstruct the sunlight from their solar array. State and local governments are very interested in promoting the expansion of solar energy and are looking for ways to protect solar access.

Many states passed solar access laws during the energy crisis of the 1970s. Currently, 34 states and approximately a dozen municipalities have some form of solar access law. In 2008, the issue of solar access made headlines when a family in California was forced by the court to cut down two redwood trees that shaded a neighbor's solar panels. The judgment was based on the Solar Shade Act (AB 2321) which California passed in 1978.

In October 2008, the U.S. Department of Energy's Solar America Board for Codes and Standards released a model solar access statute intended to encourage dialogue on the issue. As of March 2009, most of the 25 major U.S. cities enrolled in the Department of Energy's Solar America Cities program were reviewing their solar access laws.

5. Plan for the future

Your site survey should include a forecast 20 years into the future, because your solar array will be operating for at least that long. By then a five-foot spruce sapling may have grown into a monster. You should also be aware of any plans to build new buildings around your home, or any other structures that might restrict your access to sunlight.

Minimizing the effects of shading

If some shading of your solar array cannot be avoided, there are some ways to minimize its effects, including making your solar array larger to compensate for decreased inefficiency.

- **Compensate with additional panels.** If the shade occurs only during a particular part of the day, you can compensate by buying additional solar panels and angling them away from the obstruction to maximize production during that period.
- **Mount crystalline PV modules horizontally.** Crystalline PV modules should always be mounted horizontally and not vertically. Each crystalline module has two bypass diodes that are active if shading occurs. If the module is mounted horizontally and shade falls across the bottom half, only one of these bypass diodes will be active and the module will still produce up to 50 percent of its capacity. If the panel is mounted vertically, shade across the bottom will activate both bypass diodes and shut down the output of the module.
- **Change the shape of your array to avoid the shade.** Instead of a rectangular array, you might be able to arrange the solar panels in a different shape to avoid shade from a chimney or pole, or divide the array and place panels on either side of the shaded area.
- **Wire shaded modules in a separate string.** When all the solar panels are wired together, shade falling on one of the solar panels diminishes the output of all the other panels. If possible, wire the shaded module separately so the rest of the system can continue to function at optimum efficiency.
- **Use an inverter with multiple inputs.** Some inverters have a separate input for each string, so if one string is shaded the other inputs will still operate in MPP.

6. Plan the location of batteries and other components.

After deciding on a site for your solar array, you must find a suitable location for batteries, controllers, and inverters. Try to keep all the components in close proximity to each other, because long cables mean loss of power. Batteries are sensitive to moisture, extreme temperatures, and sunlight. Look for a location that is:

- Protected from precipitation and moisture.
- Out of direct sunlight.

* Ventilated to prevent the accumulation of explosive gases given off by the batteries when they are charging.
* Protected from anything that could ignite a fire.
* Secured from children and pets.
* Insulated against extreme temperatures.

Batteries are usually mounted on a rack and protected by a wire mesh cage. They should not be set directly on a concrete floor because that will lower the temperature of the batteries in extremely cold weather. If temperatures in the battery area are likely to drop below 32°F (0°C) or rise above 113°F (45°C), insulate the batteries with Styrofoam sheets around the sides and underneath the batteries.

Do not place insulation on top of the batteries as this will prevent them from ventilating properly.

Controllers and inverters should be mounted indoors, as close as possible to the batteries. They are often mounted on walls; if the inverter is large and heavy, make sure the wall is strong enough to support the weight.

7. Measure for cables.

Study the route the cables will follow from the solar array to the controller, batteries, and inverter. Measure and make a note of each distance.

After completing a site survey, you will have gathered all the important information you need to plan your system: how much power it must produce, how much sunlight is available, a rough cost estimate, and its approximate size and location. The next chapter will help you select the components that are most appropriate for your system.

Selecting the Components for Your Solar Electricity System

You will be choosing from a bewildering variety of solar products and technologies. New solar products are coming on the market every day, along with improvements on older technologies. Some products are so new that there has been little time to test them. The prices of solar components have been dropping by as much as 20 percent every year, as manufacturers expand their production capabilities and less expensive technologies are developed. By the time you read this book, there will be more new products, and prices may have dropped even further. This chapter gives you an overview of the components of a solar system and the points you should consider when putting your own system together.

The solar industry is approximately 30 years old. A contractor who has been in the business for several years should be very familiar with solar products and know which products work best in your local area. A solar contractor may be a dealer for one or more brands of solar products, or may prefer certain types of solar panels and components. If you are connecting to the grid, your local utility may recommend a standard system used by their other customers. Every installation is unique, however, because every site has unique characteristics.

Depending on the type of system you are designing, your shopping list might include:

* Solar panels
* Mounts, frames, and hardware
* Cables
* Batteries
* Controller
* Inverter
* Racks and wire cages to house the batteries
* Junction and distribution boxes
* Fuses, circuit breakers, safety disconnects
* Lightning protection
* Grounding circuit to protect the system and the people using it
* Meters and monitors
* Tracking devices to arrays facing the sun
* Wind turbine
* Back-up generator

Deciding on Your Optimum Voltage

Most solar panels and batteries are DC 12V, but only small systems can run efficiently at 12V. Solar panels and batteries are connected in series to create higher voltages for larger systems. The ideal voltage for your PV system depends on several factors.

Grid-tie system

A grid-tie system connects to the grid, so it must produce high voltages and use an inverter to convert to AC current and power your 120V and 240V household appliances.

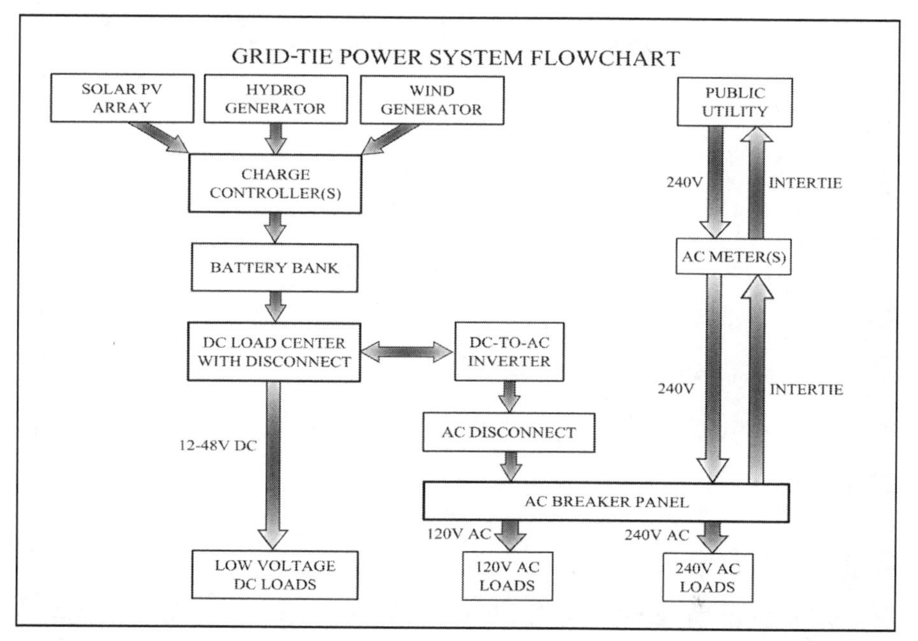

Grid-tie system with a battery fallback.

Off-grid or grid fallback systems

Off-grid or grid fallback systems typically run at 12V, 24V, or 48V.

Remember the relationship between watts, volts, and amps, described in Chapter 2: Understanding Electricity. Voltage causes current — volts are a measure of the "pressure" under which electricity flows. Current, the flow of electricity through a cable, is measured in amps (A), represented by the symbol "I." Electric power exists only when current is flowing and is measured in watts (W). Watts measure the energy that is being consumed at any given moment.

Watts (Power) = Volts × Current (Amps)

$$P = V \times I$$

Power is also equal to the square of the current multiplied by the resistance:

$$P = I^2 \times R$$

1 watt = 1 amp multiplied by 1 volt

1 amp = 1 watt divided by 1 volt

Take the example of a 60w light bulb:

Running from 120V electricity from a utility, the bulb requires 0.5 amps of current

$$W \div V = I \qquad 60 \div 120 = 0.5$$

The same bulb running from a 12V battery requires 5 amps of current

$$W \div V = I \qquad 60 \div 12 = 5$$

Running from a 24V battery, the bulb requires 2.5 amps

$$W \div V = I \qquad 60 \div 24 = 2.5$$

What difference does it make whether 5 amps or 2.5 amps of current is flowing through your system? When current flows though a cable it encounters resistance (R), which causes some voltage to be lost. Resistance is measured in ohms and can be calculated by dividing the voltage by the current.

$$R = V \div I$$

The higher your current, the more resistance it encounters. A higher voltage means lower current, lower resistance, and less loss of power in your system.

You can reduce resistance by using thicker cables, but a large-sized cable can be impractical. Thick cables are more expensive than smaller cables, and the price difference could exceed the cost of increasing the voltage of your system.

A frequent demand for high current can also reduce the life of your batteries. Battery life will be affected when the current drain exceeds 1/10 of the amp-hour rating of your battery.

When deciding to run a 12V, 24V, 48V, or 120V system, consider:

The appliances you will be running on your system:

Are you installing a PV system to operate a well pump, power lights, and a television in a weekend vacation cottage, or to provide all the electricity for a household of several people? A 12V pump that refills a water tank could run directly from a 12V solar panel without a battery, because it would not need to run at night. Conventional appliances require 120V, an inverter to step up the voltage, and back-up batteries to provide electricity when the sun is not shining. You can now purchase a whole range of appliances and electronics that run on 12V/24V but these are often smaller and more expensive than their conventional counterparts. Homeowners with off-grid systems become very conscious of how much power they are using when various appliances are running.

TIP: Weigh all the options before you decide on the voltage for your system.

Some solar experts contend that it is cheaper to install and run a 120V PV system and keep your conventional appliances than to replace all of them with new 12V/24V appliances. During your design stage, consider all the options. Do you have funds available to purchase new appliances along with the components for your PV system?

The distance between your solar array and the inverter and batteries.

To conserve power, you should always try to keep the components of your system as close together as possible. The longer the cables, the greater the resistance. A low voltage system requires thick cables. A higher voltage system can push energy more efficiently through a thinner cable. If you are placing a solar array on a pole or shed or on the ground at some distance from the house, you will need a higher voltage. A system that has solar arrays located more than 75 to 100 feet from the batteries should be at least 24V. If the thickness of the cable required to carry your current is more than ¼ inch, increase the voltage of your system.

Your budget.

Unless money is no object, there may be some trade-offs between achieving maximum efficiency and keeping your system affordable. For example, tracking devices that optimize the production of your solar panels by moving them to face the sun throughout the day could add thousands to the cost of your system. A 24V system costs less to implement than a 48V system. Inverters for 120V systems are more expensive. You might decide to switch to a propane stove or dry your clothes on a line to keep your energy demands lower and stay with a lower-voltage system.

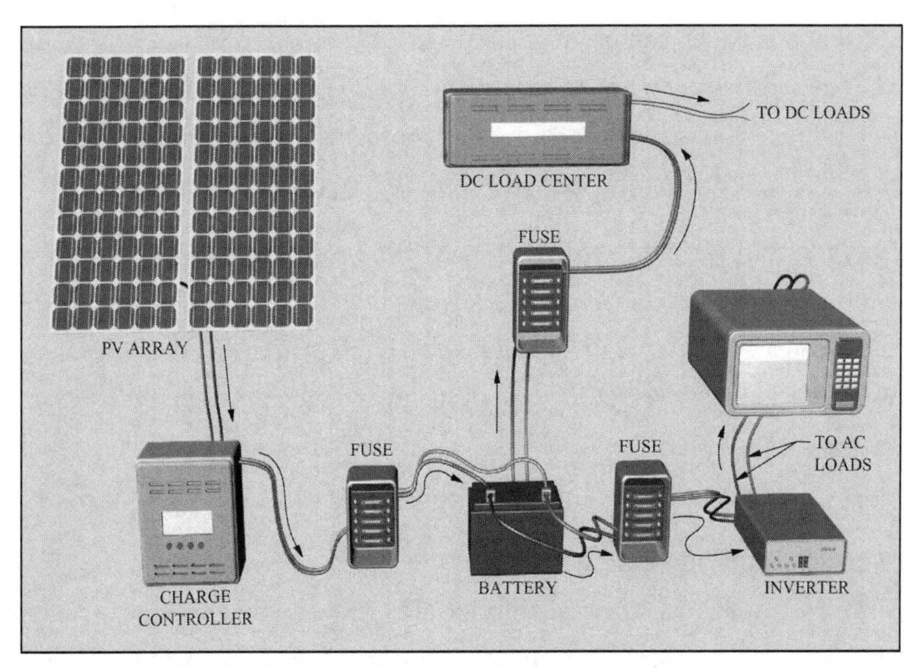

Off-grid system with AC and DC.

Solar Panels

Once you have decided on the voltage and know how many watts you need, it is time to shop for solar panels.

Three basic types of solar panels are currently used for residential installations: amorphous, polycrystalline, and monocrystalline solar panels.

Amorphous solar panels, also known as "thin film solar panels" are the cheapest to manufacture but have only 6 percent conversion efficiency. As a result, they are large and can only be used where there is no size restriction for the solar array or for a system with low power requirements (up to 200 watts). Larger systems would require so many panels that the additional expense of mounting and wiring them would make them impractical.

Amorphous panels can produce electricity in hazy and overcast conditions, and sometimes even in bright moonlight. The cost of manufacturing amorphous panels and solar films is dropping every year, and they are the main focus in developing cost-effective PV systems.

Polycrystalline solar panels are made from solar cells with a conversion efficiency of 12 to 16 percent, and a 12 volt polycrystalline panel is about 1/3 the size of an equivalent amorphous panel. Their guaranteed life expectancy is longer, about 25 years. The cost of manufacturing polycrystalline panels makes them about 50 percent more expensive than amorphous panels, but their prices are also dropping from year to year.

Monocrystalline solar panels are the most efficient panels available, with a conversion efficiency of 14 to 22 percent. They are the smallest and the most expensive solar panels, costing 20 to 30 percent more than polycrystalline panels.

String-ribbon is a relatively new manufacturing technology that combines thin film with crystalline manufacturing techniques to double the number of solar cells per pound of silicon. Its efficiency rivals monocrystalline solar panels at a lower cost.

Carbon footprints of solar panels.

A carbon footprint is the total amount of CO^2 and other greenhouse gases emitted over the full life cycle of a process or product, or over the lifetime of a person. Many homeowners want to use solar energy to reduce their carbon footprints. Clean energy has a carbon footprint just like fossil fuels. The carbon footprint of a solar panel mostly results from the manufacturing process and from its disposal. The carbon footprint of solar panels is expressed as grams of CO^2 equivalent per kilowatt-hour of electricity generated (g CO^2eq/kWh). Carbon payback is the amount of time a solar panel must be in operation to compensate for its carbon footprint. Because a solar panel must generate a certain amount of energy to replace the energy used to create it, carbon payback is shorter in climates with many hours of sunlight than in cloudy areas.

A recent study conducted by the Institute of Science in Society (ISIS) of solar panels in northern and southern Europe estimated the carbon paybacks of the three types of solar panels as:

- **Amorphous solar panels: .9 to 1.5 years**
- **Polycrystalline solar panels: 1.3 to 2.9 years**
- **Monocrystalline solar panels: 2.1 to 3.6 years**
- **String-ribbon panels: Less than one year**

Solar panels with higher conversion efficiency typically cost more than panels with lower efficiencies because the manufacturing process is more complex and expensive. A high conversion efficiency gives you more power per square foot. If you have ample space for your solar array, there is no reason why less efficient solar panels cannot do the job — you just need to cover more square feet with them. An exception would be when very high power needs require you to install large numbers of the less efficient panels — then the cost of mounting and wiring them might outweigh the price difference.

Shapes, sizes, and aesthetics

PV module front view

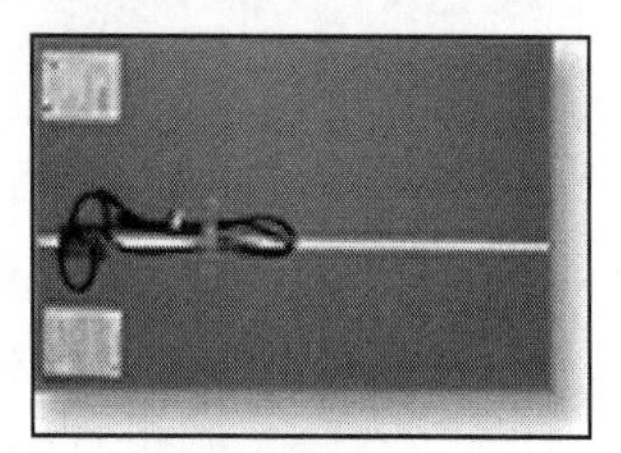

PV module back view, with connector cables

The most common solar panels are rectangular configurations of solar cells mounted in aluminum frames and covered with glass or plastic. These come in various sizes — for a small solar application a single rectangular solar panel might supply all the power you need. Less conventional shapes tend to cost more but may fit more accurately into available space or be more aesthetically pleasing. You can find triangular solar panels and panels that are integrated into architectural features such as shingles, roof tiles, and windows. Solar panels can also be installed as awnings or shades over a walkway or patio. Some panels incorporate gutters and flashing so that they can blend in with the rest of a roof. Flexible solar panels come in rolls that can be spread across a flat roof. These are not as efficient as other types of solar panels, but are inexpensive if you have enough space and do not require mounts.

Blue solar panel

Typical solar panels are blue or black. A large solar array spread over a roof can be incongruous with a carefully planned architectural design. If your solar array will be prominently visible, consider its appearance when choosing your solar panels. Architects designing new energy-efficient homes incorporate the location and "look" of the solar panels in the overall design concept.

Comparing solar panels

When you begin shopping around for solar panels, you may find yourself bewildered by the terminology and the sheer quantity of information. It is important to compare solar panels with similar attributes.

The chart below compares some typical specifications for three brands of solar panels. Once you have narrowed down your selection, a chart like this can help you to compare solar panels.

Specifications for several Kaneka, Trina, and Sharp Solar Panels

BRAND AND MODEL NAME	WATTS	AMPS	VOLTS	TOLERANCE	WEIGHT (lbs)	Warrranty	Where manufactured
Kaneka G-SA060	60	0.9	67	10/-5%	30.2	First 12 years @90% output; 25 years at 80% output	Japan, USA, and Belgium
Kaneka P-LE055	55	3.33	16.5	10/-5%	32		
Trina 175 TSM-DA01	175	5.3	36.2	0/+3	34.4	First 10 years @90% output; 25 years at 80% output	China
Trina 220, TSM-PAO5	220	7.6	29	0/+3%	43		
Trina 225, TSM-PAO5	225	7.66	29.4	0/+3%	43		
Trina 230, TSM-PAO5	230	7.78	29.8	0/+3%	43		
Trina 230, TSM-DAO5	230	7.66	30	0/+3%	43		
Sharp 142	142	7.11	20	10/-5%	32	First 10 years @90% output; 25 years at 80% output	Made in USA, Japanese-owned
Sharp NE-170UC1	175	4.9	35.4	10/-5%	35.3		
Sharp 175	175	4.95	34.8	10/-5%	38		
Sharp 180	180	5.02	35.86	10/-5%	38		
Sharp 80	80	4.67	17.3	10/-5%	19		
Sharp ND-224UC1	224	8.33	29.3	10/-5%	44		
Sharp NU-U230F3	230	8.4	30	10/-5%	44.1		
Sharp ND-U235F1	235	8.6	30	10/-5%	44.1		
Sharp NU-U240F1	240	8.65	30.1	10/-5%	44.1		

Watts — Power at STC (Standard Test Conditions): The power output of a panel when it is receiving 1000 watts per square meter of solar irradiance at 25°C.

Volts — A single PV cell produces 0.5 volts. Individual solar cells are combined in solar panels designed to produce a certain voltage — common voltages are 12V and 24V. The current produced by a solar panel varies with the amount of sunlight falling on the panel, but the voltage remains constant. Solar panels can be wired in series to create higher voltages. The rated voltage of a panel is usually higher than the voltage required from the panel to compensate for power loss — a 12V panel is typically rated 17V.

Amps — Amps is the current you can expect from the solar panel under ideal conditions. Current (I) is calculated by dividing watts by volts: $I = W \div V$.

Power Tolerance (%) — The range within which a panel will overperform or underperform its stated power in ideal conditions. A 180-watt panel with a rated power tolerance of +/-5 could output anything from 185 watts to 175 watts in full sunlight. Look for a small negative number or no negative numbers at all.

Power Warranty (Years) — The warranty on the power output of your module is based on the stated STC power minus the power tolerance. Most power warranties provide two different power guarantees: a 90 percent power output for a certain amount of time (such as ten or 12 years), and then an 80 percent power output for an extended amount of time (such as the next ten years).

In addition to the specifications in the chart above, you will probably see some other important information in the descriptions of solar panels:

Rated Power Per Sq. Ft. (Watts) — This is a measure of the watts produced per square foot of the panel. The higher the power density the less space is needed to produce a certain amount of power. A solar panel with a rated power per sq. ft. of 10 watts would give you 100 watts for every 10 square feet (10 x 10 = 100).

Conversion Efficiency (%) — Conversion efficiency tells you how much of the solar radiation received by a panel is converted to electricity. A panel with a conversion efficiency of 10 percent would create 100 watts of electricity when hit with 1,000 watts of sunlight.

Materials Warranty (Years) — In addition to a power warranty, a solar panel comes with a warranty for its materials and workmanship, usually for one to ten years.

Cell Type — This tells you what type of silicon is used to make the module's solar cells — amorphous, monocrystalline, multicrystalline, or ribbon.

Maximum Power Temperature Coefficient (% per °C) — This is the percentage that the power output of a solar panel will decrease for every degree that the temperature rises above 25°C.

Secondhand solar panels can reduce the cost of a PV system

Secondhand solar panels can be purchased very inexpensively from some solar product dealers, and are sometimes listed for sale on sites like Craigslist.com and eBay.com. Used solar panels typically become available when businesses or homeowners upgrade their systems. Some solar panel manufacturers and dealers sell returned panels, discontinued models, display panels, and "blemished" panels at a discount. Though secondhand panels can make your system much more affordable, there are some considerations:

- Used solar panels typically do not come with a warranty. "Blemished" panels are new panels with slight defects that do not affect their performance and often do come with a manufacturer's warranty.
- Most rebate programs and incentives require the professional installation of new panels with warranties.
- It is difficult to predict the life of a used solar panel. The output of new solar panels from the factory is usually 10 percent greater than the output printed on their labels. Though the output decreases over time as panels deteriorate in the sunlight, the output from an old panel can still be close to its labeled output. Solar panels that have been in storage do not deteriorate.
- Though new solar panels are guaranteed for 20 to 25 years, they often continue to function well long after that. Some panels are still in use after 40 years.
- Older panels are larger than their modern counterparts and may not be practical if space is a problem.
- Used solar panels may be damaged. Do not buy panels that are scratched, cracked, or have moisture or condensation under the glass. Other common problems are loose connections and burned-out bypass diodes. These problems can be fixed by a knowledgeable person, but it is extra effort that may not turn out well.
- The hard surface of some used panels may have turned brown over time. Check the output of these panels as they may still be operative and will not degrade any faster than clear panels.
- Test the output of used solar panels before buying. Test a panel by setting your voltmeter to DC volts, and measuring across the + and - terminals of the panel while it is exposed to sunlight. A 12-volt panel should show about 21 volts in full sun. Panels designed to be connected in sets of four (four panels in series) will show 4 to 5 volts. To determine how much current you can expect from the panels, set the voltmeter to DC amperes (high range) and connect it between the + and - terminals in full sun. For a 12-volt panel, multiply the current of a 12-volt panel by 17 volts to get watts.

Certifications and PTC ratings

In addition to STC ratings, rebate programs in California and many other states require that solar panels have PTC ratings given by a third-party testing center. PTC refers to PVUSA Test Conditions, which were developed during the 1990s to test and compare PV systems as part of the PVUSA (Photovoltaics for Utility Scale Applications) project. The PTC rating is considered a more realistic measure of output than an STC rating because the test conditions are more like "real world" solar and climatic conditions. PTC are 1,000 Watts per square meter solar irradiance, 20 degrees C air temperature, and wind speed of 1 meter per second at 10 meters above ground level. PTC ratings tend to be lower than STC ratings. Rebate programs specifying PTC ratings will be based on those lower numbers.

Nearly all electrical installations in the United States must comply with the National Electrical Code (NEC), originally developed in 1897 to protect the public from electrical hazards. In 1984, *Article 690 Solar Photovoltaic Systems* was added to the Code to address safety requirements for the installation of PV systems. It is updated regularly. The NEC applies to nearly all PV power installations, even those with voltages of less than 50 volts [720], including stand-alone and utility-interactive systems, billboards, other remote applications, floating buildings, and recreational vehicles (RVs) [90.2(A), 690].

According to the NEC, photovoltaic equipment manufacturers should build equipment to meet Underwriters Laboratories (UL) or other recognized standards and have their products tested and listed. The three most commonly encountered national testing organizations acceptable to most jurisdictions are the Underwriters Laboratories (UL), Canadian Standards Association (CSA), and ETL Testing Laboratories, Inc. (ETL).

The NEC suggests (in some cases requires), and most local inspection officials require, that PV modules be certified to UL standards in the United States, to the Canadian Electrical Code in Canadian markets, and to International Electrotechnical Commission (IEC) standards for other world markets. Most building and electrical inspectors in the United States expect to see a listing mark (UL, CSA, ETL) on electrical products used in electrical systems. In the United Kingdom, inspectors may insist that solar panels be certified by the Microgeneration Certification Scheme (MCS).

In July 2008, Underwriters Laboratories opened North America's largest commercially focused photovoltaic testing and certification facility in San Jose, California. UL's 20,000 square-foot Photovoltaic Technology Center of Excellence has 14 test chambers and two solar simulators to provide indoor and outdoor testing capabilities for evaluating photovoltaic modules and panels and a wide variety of power systems accessory equipment. This facility is intended to meet the increased demand for testing of new solar products.

For some PV manufacturers, low production rates may not justify the costs of testing and listing by UL or another laboratory. Some solar panel manufacturers claim their product specifications exceed those required by the testing organizations, but inspectors readily admit that these claims are unsubstantiated.

You may also encounter the acronym ASHRAE (American Society of Heating, Refrigerating, and Air-Conditioning Engineers) (**www.ashrae.org/aboutus**), which sets standards for heating, ventilating, air conditioning, and refrigeration and provides education and certifications for builders and designers. ASTM International, originally known as the American Society for Testing and Materials (ASTM) (**www.astm.org/ABOUT/aboutASTM.html**), develops technical standards for materials, products, systems, and services.

Brands

According to the Worldwatch Institute, the global growth rate for PV products has increased 30 percent annually during the last few years. Increasing demand and the rapid development of new technologies has spawned hundreds of new PV panel manufacturers. Be careful when looking at an unknown brand of solar panels. Cheaper crystalline solar panels may not perform as well as their brand-name counterparts. The manufacturing process for amorphous panels is simpler, and budget amorphous panels tend to be just as efficient as their more expensive counterparts.

Solar panels come with 20- or 25-year warranties. A new, little-known manufacturer may not be around in 15 years to honor a warranty. Choosing a reputable brand is an extra guarantee that you will be able to get service when you need it. The country of manufacture may also be significant — some solar contractors avoid panels made in China because manufacturing standards there are not strictly enforced.

Some well-established brands are Kyocera, Sharp, BP, Panasonic, and ClearSkies.

Final test: Use an Internet search engine

When you have narrowed down your choice of solar panels, do a quick search on the Internet for any reported problems or complaints. Type the brand name and model number in a search engine, then try some variations such as the brand name and "solar panel." Read over any complaints and comments. An isolated complaint is no cause for alarm, but multiple reports of problems or failures are a warning signal.

Putting together your solar array

Chapter 6 explained how to calculate the size of your system by doing a load analysis and compensating for system inefficiencies. You will have to buy enough solar panels to supply the watts you need. Simply add up the wattage of the solar panels until you reach a number that is slightly higher than the watts you need. Voltage is determined by the way in which the panels are wired. Solar panels can be wired in series or in parallel. To increase the voltage of your solar array to match the voltage of your system, wire solar panels in series — for example, two 10-watt 12-volt panels wired in series will create a 20-watt 24-volt array. The same two panels wired in parallel would create a 20-watt 12-volt array.

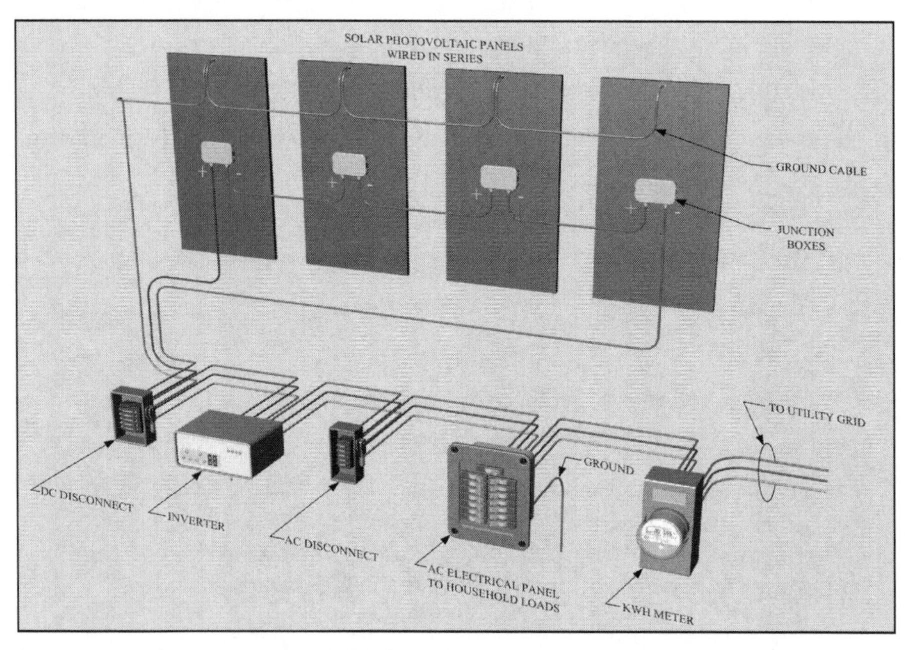

In series wiring, the negative terminal of each panel is connected to the positive terminal of the next panel. When four 10-watt panels are wired in series, the voltage increases but the wattage stays the same. To calculate the maximum amount of power the array will generate, add together the voltages of each panel.

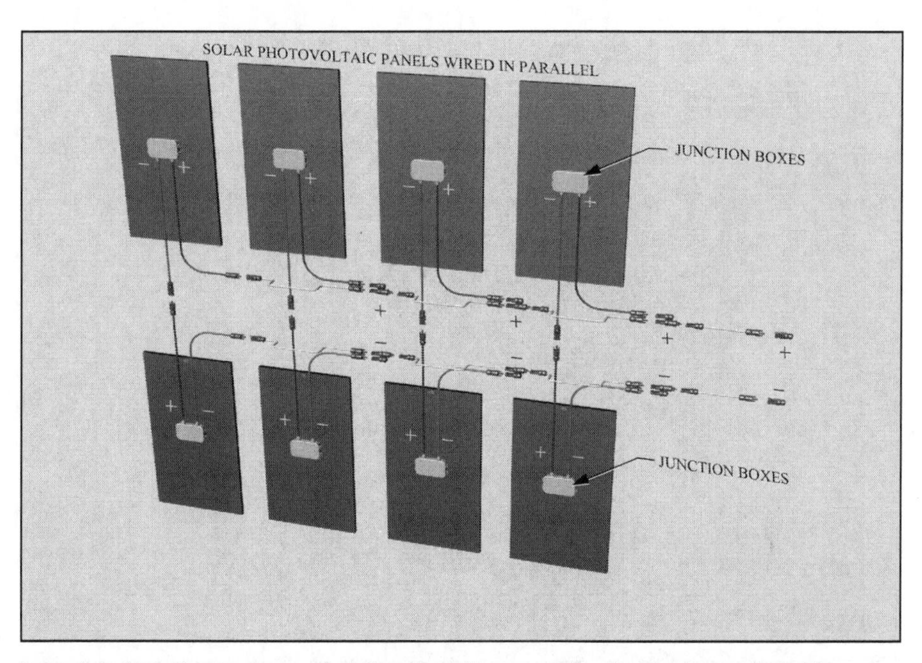

In parallel wiring, the negative terminal of each panel is connected to the negative terminal of the next panel, and the positive terminal is connected to the positive terminal. To calculate the maximum power the array can generate, average the voltage of all the panels, and add the watts. The eight 10-watt panels above wired in parallel will produce 80 watts at 12 volts.

It is possible to create two string of solar panels wired in series to achieve a particular voltage, then wire the two strings in parallel to increase watts while keeping the same voltage.

Series and Parallel Wiring

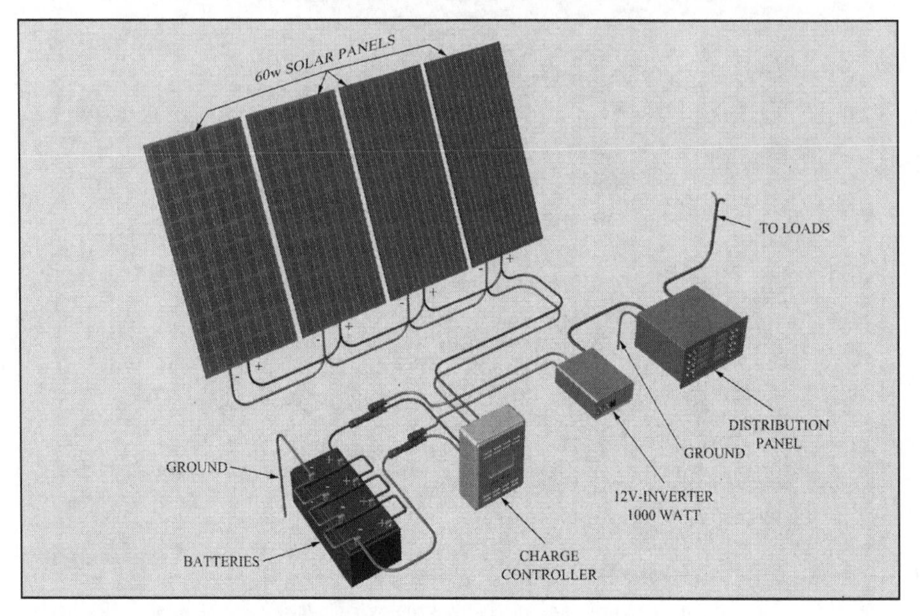

Four 60 watt panels wired in parallel will create an array that produces 240 watts at 12 volts.

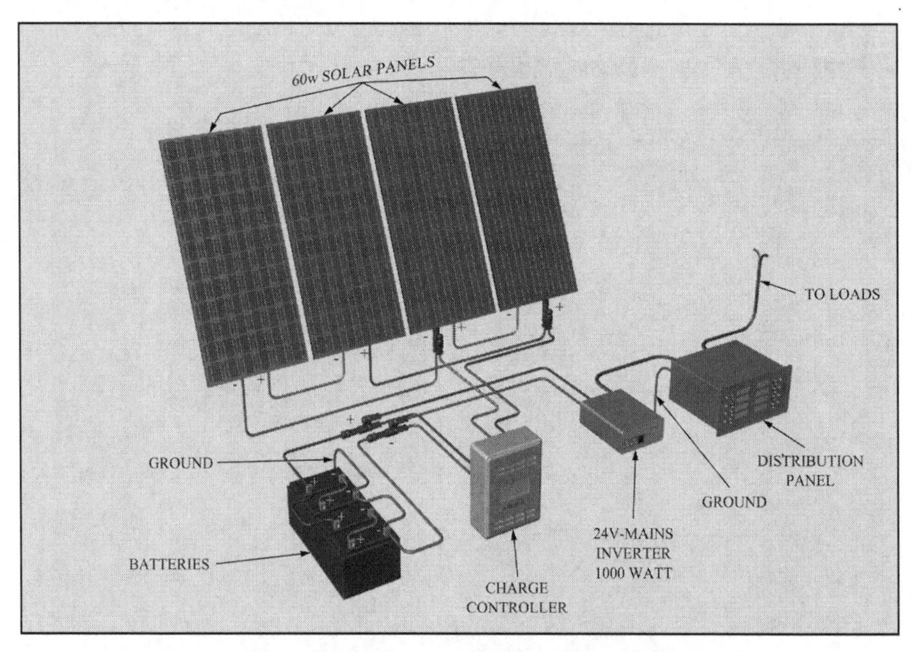

The same four panels wired in series and parallel will produce 120 watts at 24 volts.

PV experts recommend composing your array of one type and size of solar panel, rather than putting together different sizes and types. However, if two large solar panels would produce more watts than your system needs, you could use one large panel and a smaller one, wired in parallel. Create a separate string for each size of panel. You might also choose different sizes of solar panels if you are under space constraints.

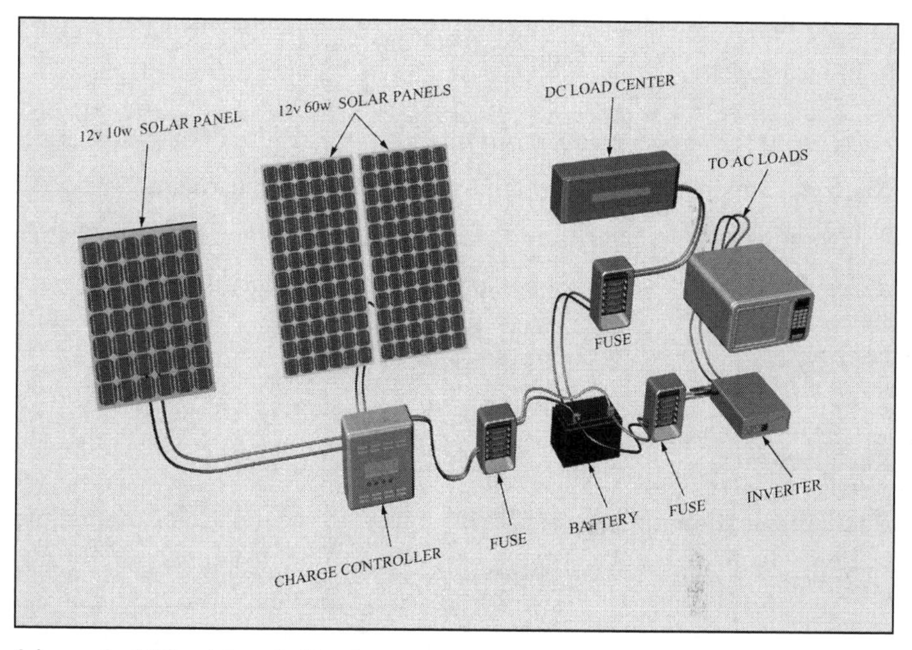

Solar panels of different sizes wired together.

Mounts and Mounting Hardware

About 95 percent of all solar installations are mounted on roofs. Your climate and roof type will determine the best mounts for your solar panels. Your solar array will remain in place for at least the next 25 years and the mounting system should be able to withstand the most extreme weather conditions in your area. Ideally, your panels should be mounted at the angle that will expose them to the optimum amount of sunlight. Most solar panel manufacturers sell mounts designed specifically for their products. A mount typically consists of fixtures that screw into the rafters underneath the roof and an adjustable rack that attaches to the fixtures and holds the panels. Installations on composition shingle roofs are the easiest and least expensive. Tile roofs are the most difficult and most expensive because the tiles break easily. Tile roofs are also prone to leaks after solar panels have been installed.

Even if you are not yet ready to put in a PV system, installing solar mounts when you are putting a new roof on your house will save money and make it easy to install solar panels in the future. It is easy to find the rafters while the roof is exposed, and any openings can be sealed against leaks.

Four types of roof mounts are common:

Rack mounts

Racks that hold multiple solar panels are fixed to the roof. Rack mounts are typically parallel to the roof surface and have the same angle of tilt.

Stand-off mounts

Stand-off mounts are adjustable frames that support the panels above the roof. The angle of the solar panels can be adjusted so that it is different from the roof angle. Stand-off mounts are often used when a house has a sunny northern exposure and too much shade on its south side. The solar panels on the northern side can be angled up so they face south. Stand-off mounts sticking several feet up from a roof can be eyesores but they provide maximum exposure for the solar panels.

Direct mount

Panels are attached directly to the roof. There is no space between the solar panel and the roof where air can circulate and cool the panels. This type of mount is cheaper because you do not have to buy racks. Flexible solar panels can be rolled across a flat roof.

Building Integrated PV (BIPV)

Solar panels are incorporated in shingles that are attached directly to the rafters, or in other architectural features such as windows or awnings.

A fixed mount on a roof is the least expensive option. A ground mount is more expensive because it requires a concrete base and a specially constructed rack, but it may be the only option if your roof cannot support the weight of a large solar array. Mounts and frames made of metal created to withstand outdoor conditions are a better choice, because even the best wood frames will not last as long as the solar panels they hold.

Solar trackers are typically created to be pole-mounted or ground-mounted. If you are using a solar tracker you must be especially aware of obstructions to the east or west, because the panels will face in those directions in the mornings and afternoons when the angle of the sun is low.

Batteries

Off-grid and grid fallback systems require batteries to store excess electricity and provide power when the sun is not shining. The cost of batteries accounts for about 10 percent of the total cost of a PV system. The batteries used for PV systems are deep-cycle batteries, meaning that they can discharge 80 percent of their power and be recharged again hundreds of times. A variety of options are available for battery back-ups.

The batteries used for PV systems are typically 12-volt lead acid batteries.

Types of batteries

There are three basic types of lead acid batteries:

Wet or "flooded" batteries

Wet batteries, also called "traction batteries," are less expensive, last longer, and perform better than sealed batteries, but they require monitoring and must be topped up with distilled water every three to six months. If you tip a wet battery, the electrolyte will spill out, creating a risk of injury or damage from the corrosive acid.

The least expensive wet batteries are golf cart batteries, which cost around $75 and are suitable for small systems of up to 800 watts. They have a life of three to five years. Marine batteries can also be used but have a shorter life (two to three years) and can only be used in banks of up to four batteries.

L16s were originally designed for use in floor sweepers and can be used for systems up to 1600 watts. They cost around $250 each, but last five to eight years.

The most expensive wet batteries cost around $700 each and last for 10 to 12 years. They can be used for systems up to 6,000 watts.

Wet batteries give off small amounts of explosive gases when charging, so they need to be placed in a ventilated area.

Absorbed Glass Mat (AGM) batteries

Also called "sealed regulated valve," "dry cell," "starved electrolyte," "non spillable," and "Valve Regulated Lead Acid" batteries, AGM batteries have the electrolyte fully contained in a very fine fiber boron-silicate glass mat. They withstand shock and vibration better than other types of batteries because the plates inside the battery are tightly packed and firmly mounted. AGM batteries do not require maintenance, will not spill corrosive liquids, and do not give off explosive gases. They have a life of four to ten years and perform best if they are not allowed to discharge more than 50 percent of their capacity before being recharged. AGM batteries do not tolerate overcharging. Leaving them unused for long periods does not diminish their performance, which makes them suitable for vacation homes.

Gel batteries (gel cells)

The electrolyte in a gel battery is a gel rather than a liquid, so the batteries do not have to be kept upright. Gel batteries do not give off explosive gases and require little maintenance. They are easily damaged by overcharging. Their lifespan is typically three to four years, and they are not suitable for systems with a power drain larger than 400 watt-hours.

NiCad batteries

Non lead-based batteries such as nickel-cadmium (NiCad) batteries are more expensive but can last a very long time if they are not discharged too deeply. A new type of nickel-cadmium battery, fiber-nickel-cadmium, has outstanding longevity at a 25 percent discharge rate. Nickel-cadmium batteries cannot be monitored in the same way as lead batteries. Because a NiCad battery has a constant output right up to the last moments before it is completely discharged, it is difficult to measure the depth of discharge. Read manufacturer's directions for maintenance of NiCad batteries.

TIP: Do not buy batteries to save for later.

It is best to use batteries as soon as you buy them. Inactivity can be extremely harmful to a battery. If you are not using the battery right away, keep it on a slight trickle charge.

Comparing batteries

Your solar contractor will recommend batteries that are appropriate for your system. Your batteries supply power at night and at times when your PV system is not supplying enough power, such as stormy winter days. Depending on your climate, your battery bank should be large enough to hold enough power to supply your system for three to five days. Multiply your daily energy requirement (as mentioned in Chapter 6) by the number of days for which you need to store back-up power to get the number of watt-hours you need and determine the size and type of batteries to purchase.

The chart below compares several brands and types of 12V batteries. You can see that the larger and heavier the battery, the greater its storage capacity.

Comparison of 12V Batteries

BATTERY NAME AND MODEL	Type	Volts	Amps	Discharge rate	Price	Weight (lbs)
Concorde PVX-1040T	AGM	12V	104AH	20 Hr		66
MK 8A31 AGM	AGM	12V	105AH	20 Hr	$244.73	69
Universal Ub121100	AGM	12V	110AH	20 Hr	$202.57	71
Concorde PVX-2120L	AGM	12V	212AH	20 Hr		138
US Battery US185	Flooded	12V	220AH	20 Hr		111
Surrette 12-Cs-11Ps	Flooded	12V	357AH	20 Hr	$1,178.12	272
Universal Ub24Gel	Gel Cell	12V	74AH	20 Hr		50
MK 8G31	Gel Cell	12V	97.6AH	20 Hr	$255.70	74

Volts — Most batteries are made up of several battery cells, each of which produces 2 volts. A 12-volt battery contains six cells. Batteries are wired together in series to produce higher voltages.

Amp hours (AH) — Amp hours is a measure of a battery's storage capacity. It tells you how many amps can be taken from the battery in a certain number of hours. AH must be looked at together with the rate of discharge — the number of hours in which that number of amps is drawn. When looking at batteries for a solar system, 100 hours (four days) would be more appropriate, but many battery descriptions give amp hours per 20 hours. Make sure you are comparing batteries with the same number of hours for the discharge rate.

To calculate how many amp hours you need, take the total number of watt-hours (three to five days of back-up power) and divide by the voltage of your system.

TIP: AGM batteries cost more but have several advantages.

AGM batteries have several advantages but cost two to three times more than similar wet batteries. They do not require any maintenance, are completely sealed against fumes and leakage, emit very little hydrogen gas when charging, rarely freeze, and will not spill even when they are broken. If your battery area is secure and well ventilated, and not subject to extreme cold, the wet batteries will serve your purpose just as well.

Cycle life

Your batteries will feed power to the PV system when it is needed and recharge whenever there is sunlight. Each time a battery is discharged and recharged, it is "cycled." After being recharged a certain number of times, a battery becomes chemically depleted and has to be replaced. The number of times a battery can be "cycled" depends on the degree to which it is discharged each time (depth of discharge, or DOD). A battery that is 80 percent discharged in each cycle can be recharged fewer times than a battery that is only 40 percent discharged. The battery manufacturer will typically supply an estimate for the number of cycles at each level of discharge that looks something like this:

Cycle life of a battery

% Depth of Discharge	10%	25%	50%	75%	100%
No. of Cycles	3200	1200	500	250	200

Your batteries will last longer if your battery bank is big enough so that it does not get heavily discharged during ordinary use. This will also give you extra back-up power when something unexpected happens, such as severe winter weather that lasts several days.

You can estimate how long your batteries will last. Suppose your battery bank is large enough to store enough back-up power for five days. During times of ordinary sunlight, your solar array will fully recharge your batteries every day, so the batteries will only discharge about 10 to 25 percent in each cycle. Assuming your system is in use 365 days a year, a battery from the chart above would last about eight years under normal weather conditions.

$$3200 \div 365 = 8.6$$

In a worst-case scenario, a winter storm lasting four days could cause your batteries to become 75 percent discharged before there is enough sunlight to recharge them again. The battery above will last for 250 cycles if it is 75 percent discharged during each cycle. Each winter storm cycle lasts four days, so 250 cycles would equal 1,000 days, or about 30 months.

4-day cycle × 250 = 1,000 days
1000 ÷ 30 = 33.3 months

The worst winter weather lasts only four months of the year, so

33.3 ÷ 4 = 8.3 years

The batteries can be expected to last about eight years before they must be replaced.

When comparing batteries and prices, look at each battery's cycle life relative to its price. If your PV system is in service for 25 years, you will need to replace batteries with an eight-year cycle life twice during that time. Buying cheaper batteries with shorter cycle lives may cut down the cost of your initial installation, but will cost you more in the long run.

Life in Float Service

Batteries have "expiry dates" — after a certain number of years they will fail, whether or not they have been in use. Life in Float Service is the maximum shelf life of a battery. When selecting batteries, you want enough power to cover emergencies, but not so much that the batteries fail before they have been through enough cycles to deplete them.

Temperature

Battery capacity is reduced when temperature goes down, and increases as temperature goes up. If your batteries will be spending several months out in the cold, you must take the reduced capacity into account when sizing your system batteries. Battery capacity is typically rated for 25°C (about 77°F). At temperatures of -22°F (-27°C) or less, battery AH capacity is

50 percent of its standard rated capacity. At freezing, battery capacity is reduced by 20 percent. At 122°F, battery capacity would be about 12 percent higher than the standard rating.

Battery voltage also varies with temperature, varying from about 2.74 volts per cell (16.4 volts) at -40°C to 2.3 volts per cell (13.8 volts) at 50°C. If your batteries will experience extreme temperature variations, your controller should have temperature compensation. Even a controller with temperature compensation will not function accurately if it is located inside at a different temperature from batteries outside, or if you have a large insulated battery bank. The inside temperature of a large battery bank may vary only 10° over 24 hours while the external temperature fluctuates 50°. When installing external temperature sensors, attach them to one of the positive plate terminals and wrap them with insulation so they maintain a temperature close to that of the battery interior.

Though high temperatures increase battery capacity, they shorten battery life. For every 15°F over 77°F, battery life is cut in half.

Size and weight

Battery specifications typically include the measurements and weight of a battery. Large lead acid batteries can weigh 60 to 100 pounds. If you will have difficulty handling a battery of that weight, select several smaller batteries instead.

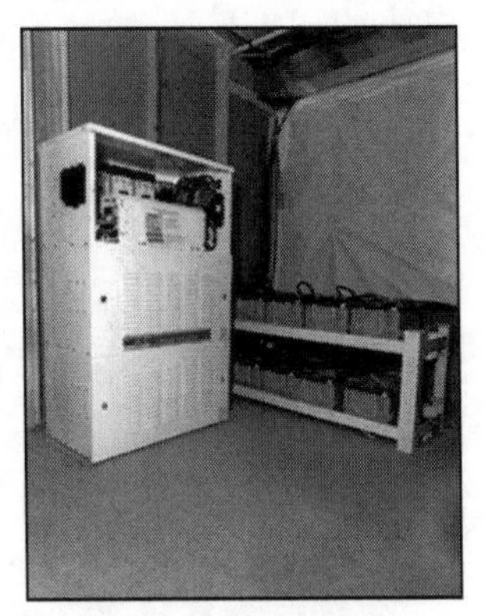

Batteries in the utility room of the Zero Energy House in Loudoun County, Virginia.

Getting the most out of your batteries

Batteries are sensitive to extreme heat and cold, and will deteriorate if they are not properly maintained. When designing your solar system, you can do several things to keep your batteries performing optimally for as long as possible.

Do not put more than four strings of batteries in parallel. If you need a battery bank with a large capacity, it is better to buy larger batteries than to wire together too many smaller ones. Too many batteries wired in parallel begin to drain power from each other.

Do not mix new batteries with old ones. If you combine new batteries with second-hand batteries that are more than six months old, the new batteries will be discharged more frequently than the old ones. The performance of the new batteries will be dragged down to the level of the old ones.

Keep the batteries at a comfortable temperature. Most batteries perform best at around 55°F, so it is best to locate them indoors and insulate them from extreme temperatures. Try to keep the batteries between 40° and 80°F. Batteries can be kept outside but should be insulated from extreme temperatures and protected from moisture.

TIP: Discharged batteries can freeze.

The sulfuric acid in a battery becomes watered down when the battery is discharged. The strong sulfuric acid in a fully charged battery will not freeze above -30°F, but a completely discharged battery could freeze at 25°F.

Maintain your batteries. Batteries need regular equalizing and maintenance and must occasionally be replaced. Your batteries should be located where you can access them easily to check the water levels of wet batteries, clean, and test them. *See Chapter 9 for information on maintaining your batteries.* Lock them away from children and pets.

Limit power use during periods of bad weather. Avoid using power unnecessarily during periods of bad weather when your solar panels are not able to charge your batteries. Your batteries are there to provide back-up power when you need it, but you can prolong battery life if you avoid discharging them too deeply. During prolonged periods of bad weather, use power sparingly and postpone non-essential tasks such as doing laundry until your solar panels are active again.

Save money with second-hand batteries

Second-hand batteries are often sold through online classified ads and by battery suppliers and golf cart dealers. They become available when battery systems are upgraded. Many of these are Uninterruptable Power Supply (UPS) batteries — they have been used as back-up power for computer systems or machinery in case of power failure — and have hardly been discharged at all. Others are from electric vehicles such as golf carts and electric cars, or rental batteries that have been returned. Though they will not last as long as new batteries, they are very inexpensive.

Do not mix old batteries with new ones — only buy second-hand when you can get enough of the used batteries for your entire battery bank. Try to find out how many cycles the batteries have gone through, and how deeply they have been discharged.

To be safe, assume that the second-hand battery will give you only half of its listed amp-hours, less for used golf cart batteries. You will probably get more than that capacity.

Second-hand batteries do not come with warranties.

Before buying, test the batteries with a battery tester if possible.

UPS batteries that have never been deeply discharged may not have a very long charge life when you buy them. To extend the charge life, connect the battery to a controller or inverter. Use a low-power device to discharge it to about 20 percent of its capacity. Recharge the battery slowly, and repeat this process two or three times to return the battery to its maximum capacity.

Wiring your battery pack

Just like your solar panels, you can wire your batteries in series to increase the voltage, and in parallel to increase the amp hours while maintaining the same voltage.

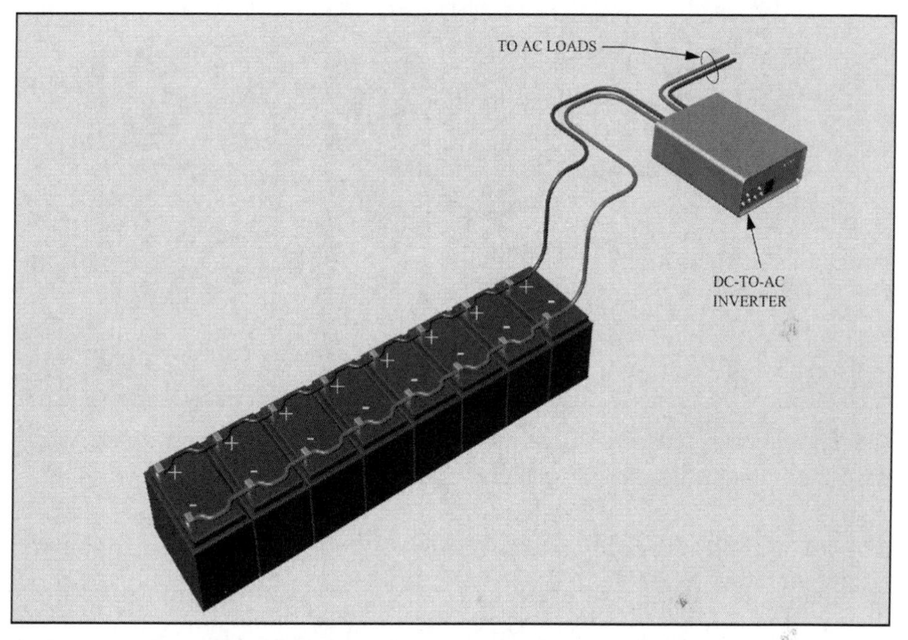

Eight 100 AH 12-volt batteries wired in parallel will give you 800 AH at 12 Volts

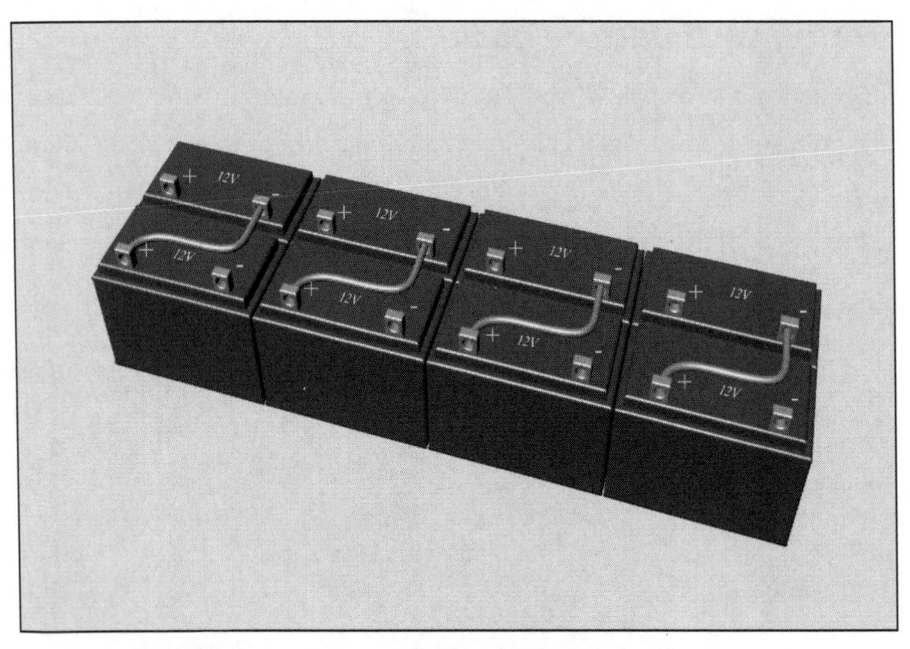

Two 100 AH 12-volt batteries wired in series will provide 100 AH at 24 Volts. (This diagram shows four sets of two batteries, with each pair of batteries wired in series.)

For example, for 200AH of storage for a system running at 24v, you will need two 12v batteries with 200AH each. As you can see from the battery comparison chart above, large batteries can be very heavy. To be able to handle the batteries easily, you might choose to buy several smaller batteries and connect them to provide enough capacity.

To create a 200AH, 24V battery pack, you can use four 100AH 12V batteries.

Wire two of the batteries in series to create a 100AH hour 24V battery unit.

Then wire two of these in parallel to give you a 200AH 24V battery pack.

The same four batteries could be wired in parallel to provide 400AH at 12V.

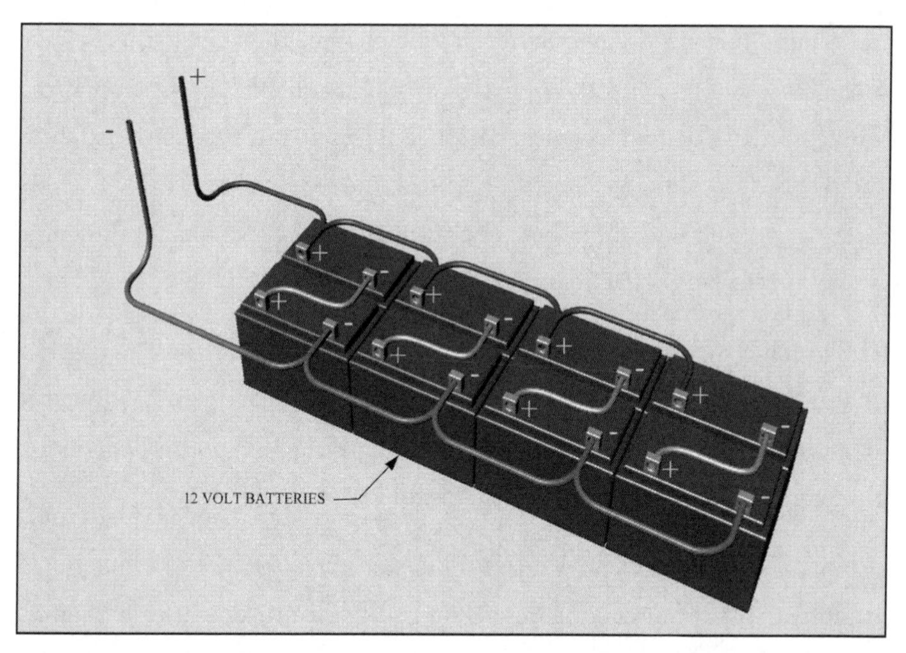

Eight 100 AH 12-volt batteries connected in series and parallel to give 400 AH at 24 Volts.

Controllers

A controller is a device that manages the power going into and coming out of your batteries. It stops them from being overcharged by your solar array and cuts off power to your appliances before the batteries become completely discharged. Controllers also shut off the circuit to your solar array at night when it is inactive to prevent "reverse flow" of current into your solar panels.

Controllers with "on/off control" regulate the flow of energy to the batteries by switching the current fully on or fully off. Controllers with "pulse width modulation" (PWM) reduce the current gradually. Both methods work well when set properly for your type of battery. A PWM controller with two-stage regulation will first hold the voltage to a safe maximum for the battery to reach full charge, then drop the voltage lower, to sustain a "finish" or "trickle" charge that tops up the battery when needed. If your system will only be used intermittently, such as for a vacation cottage that is sometimes unoccupied for days at a time, two-stage regulating is the best choice because it maintains a full charge while minimizing water loss and stress.

The voltages at which the controller switches on or off or changes the charge rate are called "set points." Some controllers allow you to adjust the set points, while others do not. The determination of set points depends on the type of battery and anticipated patterns of usage.

A battery's storage capacity varies with temperature, so the set points should be adjusted if your system is subjected to significant seasonal temperature changes. Some controllers have "temperature compensation," a feature that raises the set points when it senses a low battery temperature. If your batteries are exposed to temperature swings greater than about 30°F (17°C), compensation is essential. A controller with a built-in temperature sensor should be mounted very near to the batteries where it experiences the same temperatures. Some higher-end controllers have a remote temperature probe which attaches directly to a battery and reports its temperature to the controller.

Controllers manage the charge in each battery and ensure that each battery becomes recharged. As batteries age, their capacities and discharge rates begin to vary. A controller can detect these variations and balance the batteries as it charges them to lengthen battery life.

Controllers have a variety of tracking displays ranging from colored lights to digital displays. Most controllers have an LCD screen that shows you how much power your solar array is generating and how much charge is stored in your batteries. More sophisticated controllers allow you to compare your day-to-day usage and power generation. If you want a detailed record of your energy usage, you can purchase a separate tracking device that includes an amp-hour meter for about $200. In that case, you do not need a controller with sophisticated tracking functions.

Some controllers incorporate MPPT (maximum power point tracking), a technology that adjusts the voltage coming from the solar panels for maximum charging efficiency. *See Chapter 6, System inefficiencies for a description of how MPPT works.* Though controllers with MPPT improve the efficiency of your solar array by about 20 percent, they are more expensive. If your solar array produces 120 watts or less, it may be more cost-effective to add more solar panels to increase its output.

You can also get controllers that will start up a back-up generator when the power in your batteries is depleted.

When looking at a controller for your system, you should consider your system voltage, the current in amps supplied by your solar array, and the current of your maximum load (measured in amps). If you plan to expand your system in the future, or increase the load, get a controller with a higher current rating so you will not have to replace it. Also consider to what extent you want to monitor the output of your solar array and the charge in your batteries.

Inverters

Inverters step up the DC voltage coming from your solar panels or batteries to the 120V AC voltage used for most household appliances. You will not need an inverter for an off-grid system if all your appliances run on 12/24 volts. You will need an inverter for a grid-tie or grid fallback system, or to run 120V appliances on an off-grid system. Most homeowners with off-grid systems do not use appliances requiring 240V, but a portable step-up transformer can be used to convert 120 AC to 240 AC for an individual appliance, such as a power tool.

Inverters make up about 25 percent of the cost of an off-grid system and about 10 percent of the cost of a grid-tie system. The type and size of the inverter you need depends on the size and purpose of your PV system.

Modern inverters operate quietly and require almost no maintenance. Your solar contractor will recommend an appropriate inverter for your system. Grid-tie inverters are more expensive than inverters for off-grid systems because they are more complex. A grid-tie inverter allows you to feed AC from your PV system into the grid, and to draw power from the grid when your solar panels are not producing enough. Some inverters include battery charge controllers that manage batteries and charge them from the grid when enough solar power is not available. For grid-tie systems, your utility may require you to use a specific inverter, and for inverters to have certain certifications or ratings.

When you are buying an inverter, there are several factors to consider:

Input voltage — Each inverter is designed for a particular input voltage. Check that the input voltage of the inverter you are buying is the same as the voltage of your PV system or battery pack. If you exceed the maximum recommended input voltage, the inverter will not operate efficiently.

Maximum open circuit input voltage — The maximum output voltage from your solar array must not exceed the inverter's limit or damage can occur. Remember that the output voltage of your solar panels on an open circuit — when they are not connected to a load — is higher than their operating, rated, and maximum peak power (MPP) voltage.

Power rating — The power rating is the maximum continuous power that the inverter can supply to all the loads in your home. The power rating varies with temperature. Add up the wattages of all the appliances that might be turned on at the same time and confirm that the inverter will be able to supply enough power for all of them.

Most inverters also have a peak power rating, which is the maximum power the inverter can supply in short bursts, such as when a refrigerator motor suddenly switches on. Motors like these require extra power for a few seconds until they get started.

PV Start Voltage — This is the minimum DC voltage input required for the inverter to turn on and begin operating. Solar panels begin producing power as soon as the sun rises but may not reach this level until later in the morning. The PV start voltage is significant because a sufficient number of solar panels must be wired in each string to produce this voltage. If the manufacturer of an inverter does not provide this value in the specifications, PV system designers typically use the lower band of the Peak Power Tracking Voltage range as the inverter's minimum voltage.

Waveform — Waveform refers to the shape of the AC signal coming out of the inverter. Inverters feed DC current into several transistors using a switching mechanism that triggers a transformer to put out AC current. Less-expensive inverters put out a quasi-sine (modified sine, modified square) wave that looks like a stepped square (or pixellated) wave. This works for most standard appliances, but may not work well with electronics or appliances that have electronic heat or speed controls, or run a clock or a timer on AC. Power supplies for laptops and TVs, bread makers, chargers for power tools, some microwaves, and washing machines may not work well on quasi-sine waves. Motors may run harder, music systems may buzz, and lines appear across the picture on your TV screen.

A pure sine wave inverter puts out an AC signal identical to that supplied by a utility company. Pure sine wave inverters are more expensive than quasi-sine wave inverters, but any machine or appliance will run perfectly on their AC output. If you have only a few appliances that require pure sine wave AC, one solution is to get a large quasi-sine inverter for your general needs and a small, true sine wave inverter to use with those appliances. This would also allow you to keep small appliances such as a cordless phone running without using the large inverter full-time.

Just as with battery controllers, it is better to get an inverter that is somewhat larger than your current system or your current power needs so that you will be able to expand your system or add loads without replacing the inverter.

TIP: Be wary of lightweight, low-cost inverters made in countries with lower manufacturing standards.

High-quality manufacturing is essential to the performance of an inverter because it is subjected to intense usage and heat. Select an inverter manufactured by an American, Canadian, or European company, even if it is manufactured overseas, because these companies enforce the same manufacturing standards as in their own countries.

Inverter efficiency

Specifications for inverters include an efficiency rating which typically ranges from 85 to 95 percent. A 90 percent efficiency rating means the inverter converts 90 percent of the DC input to AC output. However, this efficiency rating is measured when the inverter is feeding power to a resistive load (a light bulb or a heater) rather than to an appliance with an electric motor such as a refrigerator or a pump. Electric motors and many appliances use 15 to 20 percent less power when they are running on pure sine AC than on quasi-sine AC. If you are running electric motors on current from your quasi-sine inverter, its efficiency will be considerably less than 90 percent. When comparing the relative efficiencies of two inverters, consider what type of appliances you will be running on them.

The efficiency of an inverter is also reduced when it is supplying less than 30 percent of its maximum power output. For example, when you run a 20-watt radio on a 1,000-watt inverter, you are really using more like 40 watts of power because the inverter is losing a certain amount of power just by operating.

Inverter efficiency varies with ambient temperature, DC input voltage, and the inverter's operating power level. The California Energy Commission (CEC) has created a "weighted" inverter test procedure that incorporates all of these variables. This weighted inverter efficiency is known as the CEC inverter efficiency.

Manufacturers' specifications for inverters often give efficiency ratings for the inverter's performance at various temperatures and output loads. Compare all of these efficiency ratings rather than just peak load efficiency.

Cooling your inverter

Inverters produce less power when they are hot. If you live in a hot climate, you need to provide sufficient ventilation to keep the inverter from overheating. Some inverters come with a heat sink or a thermostat-controlled cooling fan. Cooling fans contain moving parts and consequently have a possibility of breaking down. Heat sinks work best when there are cooling breezes and shade near where your inverter will be mounted. Inverters can be mounted indoors or outdoors, but should be placed where they are sheltered from extreme temperatures.

Diagnosis and analysis

Most inverters have some kind of diagnostic and reporting function. The simplest is a graphic display or LCD on the inverter itself. Some allow you to connect a computer directly to the inverter or use an SD card to record data. Others connect to wireless devices or have built-in servers that transmit data to a website where you can review it, or to a digital display located inside the house. The information provided can include instantaneous power, daily and lifetime energy production, photovoltaic array voltage and current utility voltage and frequency, time online "selling" to the utility today, and error or diagnostic messages. You can monitor how well your solar panels are functioning and how much AC power the inverter is putting out. You can also view stored data and see an analysis of your power production over time.

Modular inverters

Some inverters are composed of several modules, or smaller inverters wired in parallel. Each module can be independently removed for servicing. This feature is helpful if your inverter malfunctions; you will still be able to use your PV system while the faulty module is being repaired or replaced.

Micro-inverters and AC modules

The newest inverter technologies involve attaching an individual micro-inverter to each solar panel, or embedding a micro-inverter in a solar panel to create an "AC module" that outputs AC instead of DC. Micro-inverters eliminate the need for a central inverter, and even the need to wire solar panels in strings. Because each solar panel has its own inverter, the shading or malfunction of a single solar panel does not diminish the output of the whole system. Micro-inverters have built-in MPPT that optimizes the performance of each individual panel. Micro-inverters currently cost more than conventional inverters per watt (about $0.40 more), but the cost will decrease as manufacturing capabilities expand. Micro-inverters also simplify installation and save time by reducing the number of cables and connections.

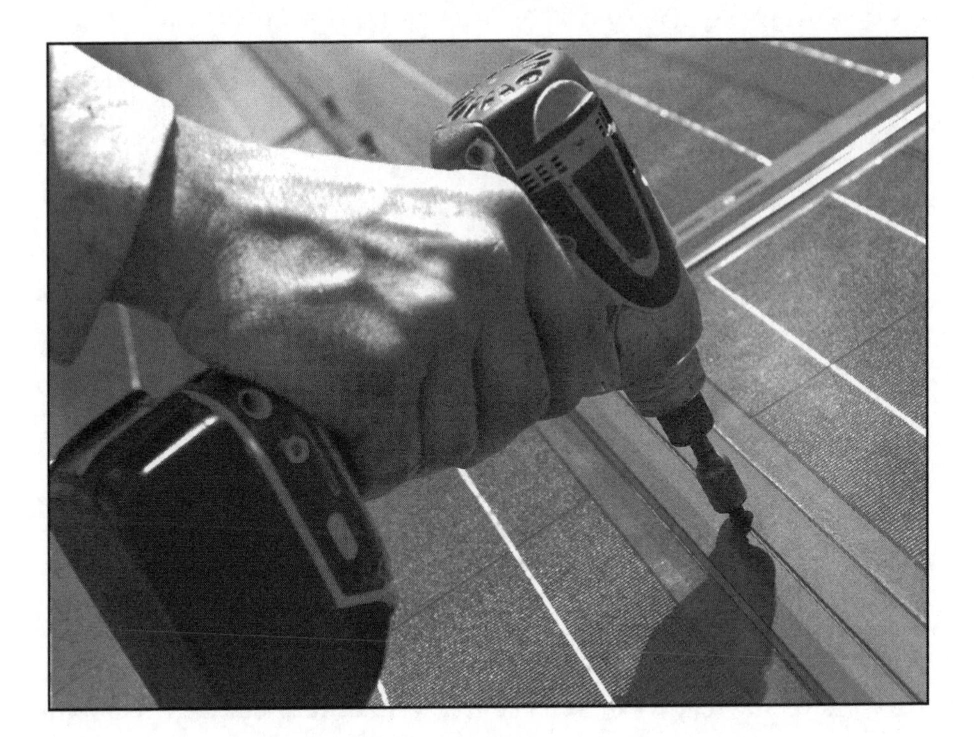

The Enphase Micro-Inverter™ System

(**www.solarpanelstore.com/solar-power.enphase_micro_inverters.html**)

Enphase Micro-Inverters are designed for use in grid-tie systems. Each Enphase Micro-Inverter is individually connected to each PV module in your array, so an individual MPPT controls each PV module. This insures that the maximum power available from each PV module is exported to the utility grid regardless of the performance of the rest of the PV modules in the array that may be affected by shading, soiling, orientation, or module mismatch. An Enphase Micro-Inverter System has three highly integrated elements: an Enphase Micro-Inverter, an Energy Management Unit (EMU), and Enlighten Web-based analytics and visualization.

Solar systems using Enphase Micro-Inverters are extremely simple to design and install. No string calculations are needed and there are no cumbersome centralized or string inverters to install. Individual PV modules may be installed in any combination of module quantity, type, age, and orientation. Each micro-inverter quickly mounts on the PV racking, directly beneath each PV module. Low-voltage DC wires connect from the PV module directly to the micro-inverter below, eliminating the risk of exposure to high-voltage DC power. The Energy Management Unit (EMU) communications gateway is installed by simply plugging it into any convenient 120V AC wall socket. The full network of Enphase Micro-Inverters will automatically begin reporting to the Enphase Enlighten web server. The Enlighten analytics software lets you know not only the immediate and historical system performance trends, but also informs you when the PV system is not performing as expected.

Installing an inverter

Your utility company must approve before you connect an inverter to the electrical utility grid. All electrical installations should be done in accordance with local electrical codes and the *National Electrical Code (NEC), ANSI/NFPA 70*. Only licensed electricians should make the connection. Rebate and incentive programs require the work to be done by professionals, and local building codes require inspections.

The inverter connects the power from your solar array with the electrical circuits in your house. All your efforts to maximize output from your carefully planned solar array and efficient use of power in your home can be sabotaged by faulty connections to your inverter. For this reason, if you are not an experienced electrician, you should consult one when installing an inverter.

TIP: Cover your solar arrays with dark opaque material before wiring them.

PV arrays produce electrical energy whenever sunlight falls on them, creating a potentially hazardous situation for anyone working on your PV system. Avoid this danger by completely covering the surface of all PV arrays with dark, opaque material before wiring them.

Solar panels wired together in series are called "strings." For maximum efficiency, only panels of the same size and type should be wired together in a string. The voltage of a string is the sum of the voltages of each panel in the series. Off-grid 12V, 24V, or 48V systems often have just one PV module per string. A grid-tie system typically has strings of several modules wired together to create the higher voltage required by the utility.

The strings coming from the solar array are joined together and connected to the main DC cable in a junction box, also called a connection box. A ground wire, or earth conductor, is also run into the junction box, which often contains a surge arrestor that directs excess voltage out of the system. Some junction boxes contain the main DC switch that controls DC input to the inverter and may contain fuses or diodes for each string. String fuses protect the wiring against overloading and are necessary for any array that includes four or more panels in a string. String diodes connected to each string ensure that if one solar module fails or malfunctions, the modules on other strings will not be affected. Without string diodes, current from the other arrays would start to backflow into the failed solar module.

TIP: String diodes also cause power loss.

Although string diodes prevent a malfunctioning solar panel from decreasing the output of the whole system, they cause a power loss of about 0.5 to 2 percent in each string because the string voltage drops when it passes through the diode. This power loss could be more or less equivalent to the power lost through a malfunctioning solar panel. Diodes can also cause problems when they fail. Evaluate your system carefully before deciding whether or not to use string diodes.

The main DC cable then carries the power from your solar array into your inverter. In a grid-tie system, the inverter may connect the solar array directly to the grid so all power generated offsets the power used in your home; or it may connect your solar array to your home electrical system so that only excess power goes to the utility grid.

In a grid-tie or grid fallback system, or an off-grid system with a back-up generator, there must be a switching mechanism to turn off the inverter when another source of AC is active.

TIP: Do not wire a generator or utility company's power directly to the AC output of any inverter.

Always wire the generator or utility company's power output to the inverter's AC input, through an external AC transfer switch if your inverter has one. Wiring the grid power or generator directly to the inverter will result in damage to the inverter and the generator.

If you have an off-grid system and only wish to run a few 120V or 240V appliances, you can create an auxiliary breaker panel to isolate the circuits for just those appliances and connect an inverter to the auxiliary panel. When you wire the inside of the house, use outlet plates of a different type or color for the AC and DC circuits so that you do not forget and plug a 12V DC appliance into a 120V AC outlet.

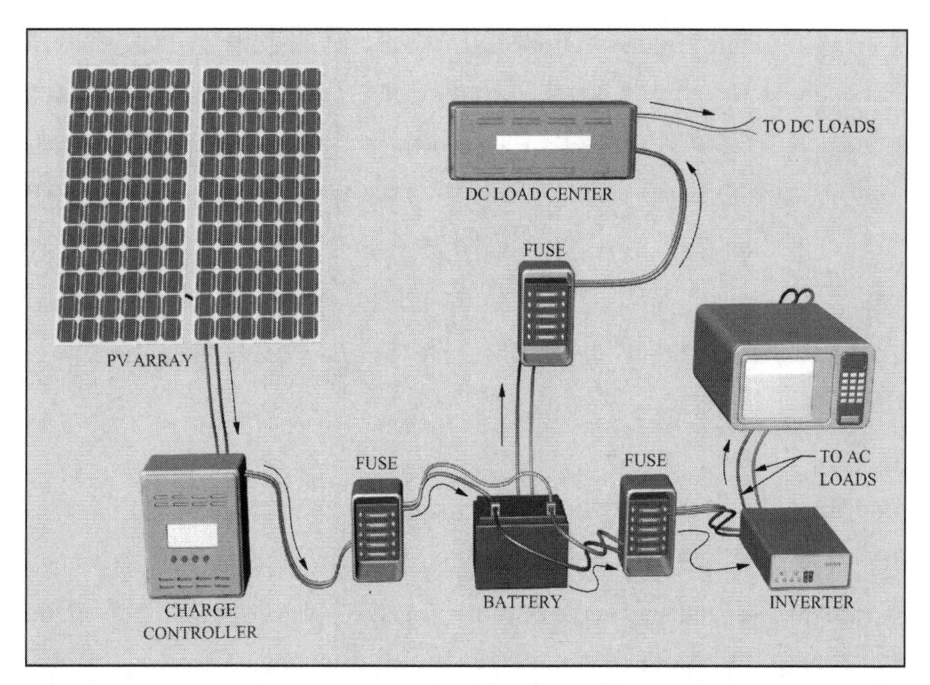

Fuses should be installed between the components of a solar system to shut off the circuit if there is a malfunction.

Cables and Wiring

It is essential to use the right cables and wires to connect your system. Electric current always encounters some resistance when flowing through a cable. You can reduce resistance by using a larger cable. If the cable you use to connect your solar array to a battery controller or inverter is too small, you will lose a large percentage of the power generated by your solar panels. Installing wrong-sized cables can sabotage your whole system. It is a good idea to use a cable that has a slightly larger capacity than you need in case you expand your system later on.

Solar equipment suppliers sell specially made cables in different lengths that have connectors already attached to the ends. These make the installation process faster and ensure that the connection is sealed against moisture. You can also purchase cabling and connectors separately and make your own cables.

When designing your system, always try to keep your system components as close together as possible to reduce the length of the cables that connect them.

Use the following method to calculate the size of the cables for your system:

Measure the distance that your cable must travel. This is your cable length. Plug the specifications of your system into this formula:

$$(L \times I \times 0.04) \div (V \div 20) = CT$$

L – Cable length in meters (1 meter = 3.3 feet)

I – Current in amps

V – System voltage

CT – Cross-sectional area of cable in meters

Most countries use the metric system, but wire sizes in the United States are still measured in inches. The table below converts American Wire Gauge sizes into millimeters.

AWG and B&S Diameter and Cross-Sectional Areas

American Wire Gauge (AWG) or B&S Number	DIAMETER		CROSS-SECTION AREA	
	Inches	mm	Square Inches	Square mm
0	0.46	11.68	0.1662	107.2
0	0.4096	10.4	0.1318	85.03
0	0.3648	9.266	0.1045	67.43
0	0.3249	8.252	0.08289	53.48
1	0.2893	7.348	0.06573	42.41
2	0.2576	6.543	0.05123	33.63
3	0.2294	5.827	0.04134	26.27
4	0.2043	5.189	0.03278	21.15
5	0.1879	4.62	0.026	16.77
6	0.162	4.115	0.02062	13.3
7	0.1443	3.665	0.01635	10.55
8	0.1285	3.264	0.01297	8.366
9	0.1144	2.906	0.01028	6.634
10	0.1019	2.588	0.008156	5.261
11	0.09074	2.305	0.006467	4.172
12	0.08081	2.053	0.005129	3.309
13	0.07196	1.828	0.004067	2.624
14	0.06408	1.628	0.003225	2.081
15	0.05707	1.45	0.002558	1.65
16	0.05082	1.291	0.002028	1.309
17	0.04526	1.15	0.001609	1.038
18	0.0403	1.024	0.001276	0.8231
19	0.03589	0.9116	0.001012	0.6527
20	0.03196	0.8118	0.000802	0.5176
21	0.02846	0.7229	0.000636	0.4105
22	0.02535	0.6439	0.000505	0.3256
23	0.02257	0.5733	0.0004	0.2582
24	0.0201	0.5105	0.000317	0.2047
25	0.0179	0.4547	0.000252	0.1624
26	0.01584	0.4049	0.0002	0.1288
27	0.0142	0.3607	0.000158	0.1021

In designing a PV system, you will need to use three types of cables:

Solar array cables — Solar array cables connect solar panels to each other and your solar array to the junction or connector box. These cables are called "array interconnects" and can be bought in specified lengths. They are used outdoors, so they must be resistant to UV rays, moisture, and extreme temperature. The NEC requires that these connector cables be located outside the building until they are near the entrance to the connector box.

Battery cables — Battery cables connect the batteries in your battery bank together, and connect your batteries to the controller and to the inverter. Ready-made battery interconnect cables can be purchased from battery suppliers. If you make up your own battery cables, be sure to use the correct battery connectors to attach them to the battery terminals.

Appliance wiring — A grid-connect system uses conventional electrical wiring. You will also use standard wiring if you are using an inverter and running your appliances on a 120V system. If you are running your household appliances on 12V or 24V, you must use larger cables to wire your house. Use the formula above to determine the size of the cables you need.

TIP: Resistance is higher when wires and cables are hot.

As wires heat up, resistance increases and more power is lost. This is not a problem over short distances, but can be significant for longer cables. If you live in a climate with very hot summers, your cables are insulated in a conduit or ganged with other cables so heat cannot dissipate easily. If the cable runs for more than 100 feet, use one wire size higher.

Ground Fault Protection Devices

Section 690.5 of the *2008 NEC* requires a PV ground-fault protection device (PVGFPD) as part of almost every PV system. Because solar panels are active anytime they are exposed to light, a short circuit or malfunction in a roof-mounted solar array could continue for hours without being detected, and could result in a dangerous house fire. GFPDs detect ground faults (malfunctions or short circuits) in PV arrays, interrupt the fault current and disconnect the malfunctioning part of the PV array, and signal the homeowner that a fault has occurred. Some inverters have built-in GFPDs. Otherwise you must purchase and install a GFPD near the inverter. PVGFPDs typically cost $180 to $300 and must be large enough to accommodate your PV system. If you plan to expand your system later, buy a larger GFPD.

A PVGFPD does not eliminate the need for branch circuit protection — each module in your solar array still needs its own fuse/breaker/disconnect to protect the wiring.

Wind Turbines

Wind turbines can charge your batteries during storms, at night — any time a breeze is blowing. The small wind turbine industry estimates that 60 percent of the United States has enough wind resources for small turbine use. In Alaska, where winter days have only a few hours of sunlight but strong winds, standard PV installations often include a wind turbine. A micro wind turbine can be installed on a pole attached to your roof or the side of your house, or on a stand-alone pole or tower. Larger wind turbines are not suitable for suburban residential areas because they typically require an acre of open space.

A wind turbine is added to your PV system calculation in a similar way to solar panels. Manufacturer's specifications tell you the output of a turbine under various wind conditions. Hybrid controllers regulate the charge coming in to your batteries from both the wind turbine and the PV system.

A solar contractor can tell you whether a wind turbine is viable in your area, and the types of turbines in use for similar PV installations. Wind turbines typically do not begin generating power until the wind speed reaches 6 to 10 miles per hour. If your area does not regularly experience winds of at least 10 to 12 mph, a wind turbine is impractical. The DOE Wind & Water Program website (**www.windpoweringamerica.gov/wind_maps.asp**) has wind maps showing wind data for each state in the United States.

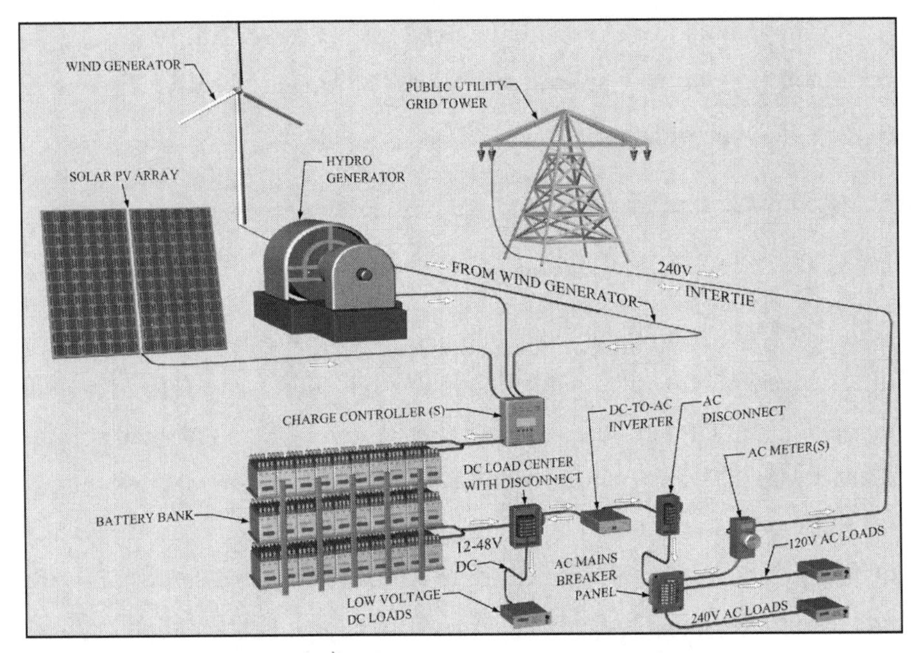

Solar system supplemented with a wind turbine and a hydro generator.

Back-up Generator

A back-up generator can run on gasoline, biodiesel, diesel, liquid propane, or natural gas and can be set up to turn on automatically when your batteries are discharged to a certain level, your utility grid fails, or your PV system is not generating the power you need. You can also turn it on manually to recharge your batteries when needed. A generator can charge your batteries during the time that a heavy load like a heater or air conditioner is drawing power from your system, and then shut off so that smaller loads like lights and clocks run off battery power. It is a waste of fuel to keep a generator running just to operate a few small loads.

Residential generators are available with outputs from 6 to 150kW. Generators can be controlled through an inverter. In a well-planned PV system, a generator will only be used 50 to 200 hours per year.

TIP: Have a generator in a grid-tie system installed by a licensed electrician.

Any interaction between an active utility line and a generator can be very dangerous. Have a licensed electrician install the generator to ensure it is done safely and correctly. During a grid power outage, a running generator could back feed electricity into the utility power lines and injure anyone servicing those lines. If your running generator is not isolated from the utility lines, serious damage to your generator and your appliances could result when the power comes back on.

Grounding

A PV system, particularly a grid-tie system, is likely to carry dangerously high voltages for up to 40 years under harsh outdoor conditions that will eventually deteriorate cables and connections. For your safety and the safety of your family, it is very important that your system be properly grounded. The NEC requires that all solar installations be connected to the earth electrically with a ground. This reduces your system's attraction for lightning, creates a safe path for lightning in the case of a strike, reduces electrical hum caused by equipment, and reduces the hazard of potential shock from the high-voltage parts of your system. If you have any doubts about your system, consult a licensed electrician or a trained inspector.

PV System Plan

If you are required to have a permit for your PV system, your local jurisdiction will probably have guidelines specifying the information that must accompany your permit application. The application will probably include a detailed line drawing showing all the components of your system and how they will be wired together. This should be accompanied by a detailed list of each component with its specifications. An inspector will verify that your plans comply with the NEC.

Even if you do not need a permit for your system, you should create a detailed wiring diagram. If possible, consult a solar expert before finalizing your plan. It is much easier to make changes in the design stage than later on when you have already purchased expensive equipment and supplies or installed the wrong size inverter or controller.

Rebates, Subsidies, Tax Credits, and Loans

Nothing can alter the fact that electricity produced by a PV system costs more than electricity purchased from a utility. At the same time, national and state governments are eager to implement the use of solar power as an alternative to energy supplied by burning fossil fuels. Governments are motivated to promote solar power because of a looming energy shortage, the urgent need to reduce greenhouse gas emissions, and political wrangling over fossil fuel resources. As a consequence, numerous government programs offer subsidies and tax credits to offset the cost of installing a PV system. To make residential solar systems truly viable, the federal government passed the Public Utility Regulatory Policy Act (PURPA) in 1978 mandating local utilities to pay their customers for excess electricity generated by their PV systems.

Governments believe that subsidizing residential solar installations will gradually bring prices down by stimulating manufacturers to expand their facilities. A strong demand for solar technology motivates research and the development of cheaper solar energy products. Prices of solar energy products have dropped considerably over the last few years. According to a 2009 study by Lawrence Berkeley National Laboratory, prices of solar panels declined 31 percent from 1998 to 2008 because of lower manufacturing and installation costs and state and local subsidies. Subsidies are directly linked to increases in sales of solar products; domestic sales of solar panels in Japan doubled in 2009 after the government introduced a subsidy for homeowners installing PV systems.

This chapter covers the various types of financial incentives for purchasing solar systems, as well as loan options and selling power to your utility. It is difficult to keep up with all the tax incentives, rebates, and subsidies because new ones are constantly being introduced and old ones are expiring. Get the latest information from your solar contractor, your tax advisor, and your local government. You can also find updates on the websites mentioned below.

You will not qualify for rebates and subsidies unless you comply with strict requirements such as having the installation done by a licensed and/or certified professional, purchasing equipment rated according to official standards, and having your installation inspected. Electrical installations must conform to local building codes and the NEC. Do-it-yourself projects may not be eligible for tax credits or subsidies. Before embarking on your solar project, study the requirements carefully. Eligibility for financial benefits may influence your decision to hire a solar installer and your choice of equipment, and should certainly be part of your financial calculations.

Several types of loans are available to finance solar installations and improve the energy efficiency of your home. Solar equipment dealers and installers often collaborate with specific lenders and can assist with the application process. Subsidized loan programs are available in certain counties and states. Because financing charges are included in the overall cost of your system, it is important to understand your loan options.

Many rebate programs and subsidies use money that trickles down from the federal government, but they are administered on a state and local level and may be specific to your county or jurisdiction. Your address and in some cases your income will determine your eligibility for these benefits. This chapter provides an overview of the types of financial benefits available and directs you to resources where you can get more specific information.

Federal Tax Credits

A tax credit reimburses part of the cost of a solar system by reducing the amount of personal income tax you must pay. It is different from a tax deduction, in which you subtract itemized amounts from your taxable income. A tax credit is subtracted from the tax you owe after all deductions have been taken. It increases your tax refund or decreases the amount of tax you owe.

The Residential Renewable Energy Tax Credit

The Residential Renewable Energy Tax Credit reimburses homeowners for 30 percent of the cost of installing a solar electric or solar water heating system at their residences.

The federal Energy Policy Act of 2005 first established a 30 percent tax credit (up to $2,000) for the purchase and installation of residential solar electric and solar water heating systems, and a 30 percent tax credit (up to $500 per 0.5 kilowatt) for fuel cells. These tax credits were scheduled to expire at the end of 2007, but were extended through December 31, 2008, by the Tax Relief and Health Care Act of 2006. They were extended again, until December 31, 2016, by The Energy Improvement and Extension Act of 2008, and a tax credit was added for small wind-energy systems and geothermal heat pump systems. In February 2009, The American Recovery and Reinvestment Act of 2009 (H.R. 1: Div. B, Sec. 1122, p. 46) removed the maximum credit amount limit for all eligible technologies (except fuel cells) placed in service after 2008. The American Recovery and Reinvestment Act of 2009 removed a previous limitation on the use of the credit for eligible projects that were also supported by "subsidized energy financing." This restriction does not apply to projects placed in service after December 31, 2008. The Residential Renewable Energy Tax Credit is regulated by the Internal Revenue Service (IRS). Here are the rules:

* The PV system must serve a dwelling unit located in the United States that is used as a residence by the taxpayer. The home served by the system does not have to be the taxpayer's principal residence.
* Eligible expenditures include purchase of equipment, labor costs for onsite preparation, assembly or original system installation, and piping or wiring to interconnect a system to the home.
* Expenditures on equipment are treated as made when the installation is completed.
* Systems must be placed in service on or after January 1, 2006, and on or before December 31, 2016. If the installation is on a new home, the "placed in service" date is the date of occupancy by the homeowner.
* If the federal tax credit exceeds the homeowner's tax liability for that year, the excess amount may be carried forward to the

succeeding taxable year. The excess credit can be carried forward until 2016 — after that, it is unclear whether the unused tax credit can be carried forward.

* There is no maximum credit for systems placed in service after 2008. The maximum credit is $2,000 for systems placed in service before January 1, 2009.

There are some additional requirements for solar water heaters and wind generators:

Solar water-heaters

* Equipment must be certified for performance by the Solar Rating Certification Corporation (SRCC) or a comparable entity endorsed by the government of the state in which the property is installed.
* At least half the energy used to heat the home's water must be from solar.
* The tax credit does not apply to solar water heating for swimming pools or hot tubs.

Small wind-energy property

* There is no maximum credit for systems placed in service after 2008. The maximum credit is $500 per half kilowatt, not to exceed $4,000, for systems placed in service in 2008.

To claim the tax credit you must fill out *IRS Form 5695: Residential Energy Credits* (**www.irs.gov/pub/irs-pdf/f5695.pdf**) and include it in your income tax return. Keep your receipts for the equipment and labor, and copies of Manufacturer's Certification Statements for each component. A Manufacturer's Certification Statement is a signed statement from the manufacturer certifying that the product or component qualifies for the tax credit. These certifications should be available on manufacturers' websites. If not, call the manufacturer.

TIP: The Residential Renewable Energy Tax Credit is not limited by the AMT.

The alternative minimum tax (or AMT) is an alternative set of rules for calculating your income tax that determine the minimum amount of tax that someone with your income should be required to pay. If your regular income tax falls below this minimum, you must pay the AMT instead. If you pay AMT, you can still receive the 30 percent tax credit for geothermal, solar, and wind installations.

A homeowner who receives a rebate from a local utility for a solar installation does not have to report the rebate as taxable income. The rebate should be subtracted from the cost of the solar system before calculating the 30 percent tax credit — the credit is for 30 percent of the net amount paid for the system. A commercial enterprise has to report rebates as taxable income and would calculate the commercial solar tax credit or Treasury cash grant based on the original cost of the system before the rebate.

The Residential Energy Efficient Property Credit

Chapter 4 discussed ways to make your home more energy efficient by caulking and sealing air leaks, adding insulation and replacing old appliances with new energy-efficient ones. The Residential Energy Efficient Property Credit reimburses you for 30 percent of the cost, up to $1,500, of products that make your home more energy efficient. Eligible products include water heaters, furnaces, boilers, heat pumps, central air conditioners, building insulation, windows, doors, roofs, circulating fans used in a qualifying furnace, and stoves that use qualified biomass fuel.

This credit cannot exceed $1,500 for 2009 and 2010 combined, and it is set to expire on December 31, 2010, unless Congress extends it.

The rules for this credit are slightly different:

- The home must be your primary residence.
- Equipment must be new and in compliance with all applicable performance and safety standards as described in tax code. Consult the f ederal Energy Star® website (**www.energystar.gov/index.cfm?c=tax_credits.tx_index**) for details.
- The tax credit does not cover labor or installation costs for insulation, doors, and roofs, only the cost of materials. Contractors should supply you with an itemized bill that separates the cost of labor and the cost of the products used.

TIP: Ask your contractor ahead of time for an itemized bill.

Tell your contractor before the work is started that you wish to apply for the tax credit and need an itemized bill. This will ensure that the contractor provides you with correct information and documentation.

Some of the items eligible for the tax credit are:

- "Metal roofs with appropriate pigmented coatings" and "asphalt roofs with appropriate cooling granules" that also meet Energy Star® requirements.
- Bulk insulation products such as batts, rolls, blow-in fibers, rigid boards, expanding spray, and pour-in-place.

* Products that reduce air leaks including weatherstripping, spray foam air seal in a can, caulk designed to air seal, and house wrap.
* HVAC units that meet Energy Star® standards.
* Advanced Main Air Circulating Fans for furnaces.
* Heat pumps.
* Gas, oil, or propane water heaters.
* Biomass stoves.

Not all products with an Energy Star® rating qualify for the tax credit.

The Residential Energy Efficient Property Credit is also claimed on *IRS Form 5695*. In 2010, unless Congress changes the law, you cannot receive this credit if you are paying AMT.

Commercial Energy Tax Credits and Grants

Several programs offer grants or tax credits to businesses, local governments, commercial builders, agricultural enterprises, and other entities that install PV systems to supply energy for themselves or for the local community.

These include:

* Renewable Energy Production Incentive (REPI)
* Qualifying Advanced Energy Manufacturing Investment Tax Credit
* Energy-Efficient Appliance Manufacturing Tax Credit
* Tribal Energy Program Grant
* U.S. Department of Treasury — Renewable Energy Grants
* USDA — High Energy Cost Grant Program
* USDA — Rural Energy for America Program (REAP) Grants

* Business Energy Investment Tax Credit (ITC)
* Energy-Efficient New Homes Tax Credit for Home Builders
* Renewable Electricity Production Tax Credit (PTC)
* Residential Energy Conservation Subsidy Exclusion (Corporate)
* Energy-Efficient Commercial Buildings Tax Deduction
* Modified Accelerated Cost-Recovery System (MACRS) + Bonus Depreciation (2008-2010)

You can find more information about commercial solar energy incentives on the DSIRE website (**www.dsireusa.org**) or the IRS website (**www.irs.gov**).

State and Local Subsidies

Many states also offer state tax credits to offset the cost of solar installations or subsidies funded by taxpayer dollars or federal stimulus money. According to Amy Heinemann, a policy analyst at the Database of State Incentives for Renewables and Efficiency, more than half the states in the United States and Washington D.C. offer enough incentives to cut the costs of installing a PV system by 40 percent or more. The price you pay for a PV system could largely be determined by where you live. In Arkansas, a state that has no solar subsidy, a homeowner could pay $35,000 for a system that would cost $2,625 in New Jersey. With an annual savings of $492.47 on utility bills, it would take more than 71 years to recoup the cost of the installation.

Recession causing some states to drop or reduce subsidies

In 2008, New Jersey reduced its subsidy by half. In May 2010, Massachusetts temporarily suspended it for several months. Colorado, Maryland, Connecticut, Minnesota, Arizona, Florida, Pennsylvania, California, and Delaware also reduced their subsidies. One reason is that so many homeowners have taken advantage of the subsidies. A decrease in new home construction during the recession may have contributed to this popularity by pushing solar contractors to market their solar products more aggressively. Falling prices of solar products and increases in the amounts paid by utility companies for electricity produced by residential PV systems will compensate somewhat for the loss of subsidies.

Some local utilities also offer subsidies or rebates to customers who install grid-tie PV systems.

Property tax exemptions

Installation of a PV system can add thousands of dollars to the resale value of your home. According to experts, the ratio of increased home value to annual electricity bill savings is 20:1. A reduction of $1,000 a year in your utility bill could mean a $20,000 increase in the value of your house. A higher appraised home value would ordinarily mean an increase in your property taxes. As an incentive to homeowners who install PV systems, 17 states, including Connecticut, New Jersey, and New York, offer a 100 percent property tax exemption indefinitely for solar photovoltaic investments. Twelve other states provide a limited period of full property tax exemption, and others offer some type of partial property tax exemption. Twenty-three states give sales tax exemptions for the purchase of solar products.

The North Carolina Solar Center at North Carolina State University (NCSU) and the Interstate Renewable Energy Council (IREC) maintain an online database of federal, state, and local renewable energy incentives called Database of State Incentives for Renewables & Efficiency (DSIRE) (**www.dsireusa.org**). It is funded by the U.S. Department of Energy's Office of Energy Efficiency and Renewable Energy (EERE) and is regularly updated.

Energy Matters, a California solar and wind energy leader, maintains a directory of state and local incentives on Solar Estimate.org (**www.solar-estimate.org**) along with free estimators for solar and wind systems.

Your solar installer is likely to have the most up-to-date information on the rebates and incentives available to you, and will know how to make sure you are eligible for them.

Rebates

A rebate is cash given back to you when you make a purchase. A variety of rebates are available to reduce the cost to a homeowner of installing a PV system. Rebates are offered by state and local governments, local utility companies, and manufacturers of solar equipment. Rebate programs are intended to reduce the energy burden on local utilities and decrease the use of fossil fuels by encouraging local homeowners to install solar systems. State and local rebate programs might be funded by federal stimulus money, state taxes, or a small monthly surcharge on all utility customers.

Manufacturers sometimes offer rebates as an incentive to buy their product instead of a competitor's. Some manufacturer rebates are part of government incentive programs to encourage the development of solar industries. A manufacturer's rebate may be paid directly to the solar contractor who installs your system, and the savings passed on to you as a lower bill. State and local programs may hand you a check to reimburse you for all or some of the cost of your PV system after you have paid in full. Utility company rebates may be in the form of reductions on your monthly energy bills.

To get a rebate for purchasing a computer at an electronics store, all you have to do is fill out a form and mail in a receipt and the UPC label from the package. Applying for a rebate on a solar energy installation can be very complicated and usually involves the cooperation of your solar installer and local inspection agency. As an example, this is the procedure for getting a rebate from the Austin Energy Solar PV Rebate Program:

1. You fill out a solar rebate participation request form on the Internet or mail it in to Austin Energy.

2. A solar inspector conducts a preliminary site survey using aerial photographs. Deed restrictions must not prohibit the installation of solar photovoltaics on your property. Your roof must be new or in good condition. The roof where your solar panels will be located must meet the requirements of AE's Solar Access Guidelines. The array must be on a south-facing roof within a certain range of tilt angle and azimuth. Shading on the site must be within acceptable limits. *See TSRF in Chapter 6.*

3. If you pass the preliminary inspection you are given an information packet including a rebate application, REC agreement, solar program guidelines, a list of registered solar contractors, and a list of energy efficiency requirements.

4. You select one of the registered solar contractors who evaluates your electrical system, fills out your rebate application (for you to sign), and presents you with a price quotation and a PVWATTS analysis. The solar contractor then submits the REC Agreement, rebate application, PVWATTS analysis, and a detailed plan of the system layout to AE.

5. A solar inspector evaluates your proposal. You are then informed if the proposal does not comply with requirements, or if funding for the rebate program is already depleted. If you do qualify for the rebate, your house is pre-screened to verify that it meets the energy efficiency standards. These include:

Existing residential homes constructed prior to January 1, 2009 must, within the past ten years, have done one of the following:

1. Completed the following recommended energy efficiency measures through the Austin Energy Home Performance with Energy Star® program, or met the minimum efficiency standards of the Energy Conservation Audit and Disclosure Ordinance (ECAD) Audit:

* All sun-facing windows and glass doors receiving at least one hour of direct sunlight in the summer and on 40 percent or more of the glass area must have Solar Screens or window treatments.
* Attic insulation shall be no less than a rated value of R-22.
* Home comprehensive air seal testing must be performed in addition to the ECAD audit. Ideally, the house should be between 0.35 and 0.45 Air Changes/Hour (ACHn) and should have mechanical ventilation if it is not.
* Air Duct System performance testing must prove that there is less than 10 percent leakage.

OR 2. Received energy-efficiency improvements through the Austin Energy Free Home Improvements program for customers with low to moderate incomes.

OR 3. Received an Austin Energy Green Building 3-Star Rating or better using version 8.0, or an Austin Energy Green Building 5-Star rating.

* Home Water Heating Systems must comply with AE requirements.

Newly built homes must meet the following standards:

* Texas Climate Vision code compliance calculated performance must exceed 10 percent over current Austin Energy Code before water heater efficiency is considered. (**http://tcv.tamu.edu**)
* Home Water Heating Systems must comply with AE requirements.

If these conditions are not met, the rebate application is placed on hold until the necessary changes have been made and verified.

6. If the rebate application is approved, AE assesses its budget for available funds and issues a Letter of Intent (LOI). This may require the approval of the City Council. If funds are not available, you are informed and put on a waiting list.
7. Your solar contractor pulls permits and proceeds with the installation in accordance with solar access guidelines and solar rebate guidelines. When the installation is complete, the contractor submits a long list of documents, including a paid invoice, warranties, lists of components with model names and serial numbers, specifications, wiring diagrams, an updated PVWATTS analysis, and your check request.
8. The contractor contacts AE and schedules an inspection. A field inspector comes and examines the entire system, verifying that all the documentation is correct and the work is done according to specifications.
9. The field inspector completes and submits an inspection form. You receive your rebate check in the mail. The AE residential solar rebate is $2.50 per watt, with a maximum rebate of $50,000. Annual rebate amounts are limited to $15,000 per site.

As you can see, the procedure is lengthy and you must fulfill many requirements. Austin Energy does not want to pay for a PV system that is inefficient in any way. You might have to spend even more money to make your home energy efficient so you can qualify for the rebate. This is not a do-it-yourself project. It is worth the effort, though. If you receive the rebate, your PV system will be paid for in a very short time and you will have free electricity for the next 20 years.

Rebates are typically distributed on a first-come first-serve basis. Homeowners who qualify for rebates may not receive them if the funding runs out. In 2010, homeowners in many states did not get the rebates they expected because funding was not available and because the programs were so popular. If you are put on a waiting list for a rebate, you will have to decide whether to go ahead and install the PV system at your own expense, or wait until funding becomes available.

Getting a Loan to Finance Your System

Low-interest loans to finance PV installations are available through many utility companies — contact your local utility or look on their website to see if they have such a program. Many solar contractors work with preferred lenders and help you with documentation for loan applications.

Federal loan programs

Several federal programs offer low-interest loans for energy-efficient home improvements, photovoltaic systems, and solar water heaters:

Fannie Mae (**www.fanniemae.com**) partners with utility companies to offers ten-year unsecured loans of up to $15,000 for the purchase of grid-tie or off-grid PV systems, and solar water and space heating systems. The Residential Energy Efficiency Improvement Loan program provides a below-market interest rate and promotes a bundled approach to efficiency upgrades.

The **Federal Home Loan Mortgage Corporation (FHLMC)**, also known as **"Freddie Mac,"** (**www.freddiemac.com**) is a secondary mortgage lender; it purchases mortgages from lenders, packages them as securities, and sells the securities to institutional investors such as insurance companies and pension funds. Freddie Mac provides specific criteria for energy efficient mortgages (EEMs) that it is willing to buy on the secondary mortgage market. Interest rates for these mortgages are either fixed at market rates or variable at prime rate plus 2 percent.

The **U.S. Department of Agriculture (USDA)** (**www.usda.gov**) has a leveraged loan program for rural borrowers that could provide rural homebuyers with an opportunity to borrow money for energy efficient equipment. It also provides loans to rural service utilities and rural development borrowers that loan homeowners in rural and remote areas money to install PV or solar thermal systems.

HOME Investment Partnership Program provides more than $1 billion a year to state and local governments for investment in long-term affordable housing for lower income families. This program can be used for housing rehabilitation that includes energy conservation. Joint ventures by state and local governments, public utility companies, and nonprofit providers are encouraged.

Several **U.S. Department of Housing and Urban Development (HUD)** programs can help with financing solar energy systems. FHA mortgage insurance is available for solar energy systems in three ways:

1. The Energy Efficient Mortgage Program (EEM) for new and existing 1- to 4-unit properties. Energy improvements must be

identified with a home energy rating and may not be valued at more than 5 percent of the property value, up to $8,000. An EEM can be used in conjunction with other FHA loans.

2. Mortgage Increase for Solar Systems. In 1978 Congress authorized FHA to exceed by 20 percent the maximum loan limit to allow for the installation of solar heating and domestic hot water systems. This authorization is now being applied to PV systems. These loans require that the home have a 100 percent operational conventional backup system.

3. Title I Property Improvement Mortgage Insurance enables lenders to make property improvement loans to creditworthy borrowers with little or no equity in their homes. The maximum loan is $25,000 for single-family homes. No energy efficiency calculations are required.

Veterans Administration (VA) loans may also be used to improve a home by installing energy-related features such as solar heating and cooling systems, water heaters, insulation, weatherstripping and caulking, storm windows, storm doors, or other energy-efficient improvements approved by the lender and VA. These features may be added with the purchase of an existing dwelling or by refinancing a home owned and occupied by the veteran. A loan can be increased up to $3,000 based on documented costs or up to $6,000 if the increase in the mortgage payment is offset by the expected reduction in utility costs. A refinancing loan may not exceed 90 percent of the appraised value plus the costs of the improvements. A veteran may refinance an existing VA loan to retrofit a home with energy efficient measures. In new construction, a photovoltaic or solar thermal system can be included in the sale price of the home.

The **Environmental Protection Agency (EPA)** has an Energy Star® Financing Program that works with lenders to provide special financing for buyers of Energy Star®-rated homes. The Energy Star® Homes Program encourages builders to construct homes that are 30 percent more energy efficient than homes built to the model energy code. Some Energy Star®-rated builders are including solar thermal and photovoltaic systems. Currently, three national lenders and several regional lenders are offering Energy Star® mortgages that allow homebuyers to purchase homes with mortgages 10% to 24% higher than they would have qualified for if the home were not Energy Star®-rated.

Home equity loans

Banks know that adding a PV system to your home will increase its resale value, and make your monthly mortgage payments more affordable by reducing your electricity bill. They are often willing to give favorable interest rates on a loan that pays for a PV system. If you are considering buying a PV system, you probably have some equity in your home. Home equity loans have low interest rates and an added financial benefit: The interest you pay is considered mortgage interest and is deductible from your taxable income.

Selling Power to Your Utility

The Federal Public Utility Regulatory Policy Act of 1978 (PURPA) allows businesses and individuals to sell excess electricity generated by solar or wind systems to their local utilities at avoided cost. Avoided cost is the utility's wholesale cost to produce electricity, and is about ¼ of the retail price. Utilities that buy excess electricity at wholesale prices install a second meter at your home to measure the outgoing power. Net metering, a program that pays you retail prices for your electricity, is now offered in more than 35 states.

Net metering

Your electricity meter literally "runs backward" in net metering. When the sun is shining brightly and your PV system produces more power than you are using, the excess electricity is fed to your utility grid. This excess electricity offsets the electricity you use from the grid when your PV system is not generating enough power, so that you end up selling electricity for the same price you would pay for it. With net metering, you will recoup the cost of installing a PV system much more rapidly.

Time of Use (TOU) price structures charge more for electricity used during peak hours. In the summer, solar systems typically produce the most electricity during peak hours, the hottest time when air conditioners are running in nearly every building. If your utility has TOU, you will earn even more credit because you will be buying at the lower off-peak prices and selling your electricity at higher peak-hour prices.

TIP: Net metering gives you an added financial bonus.

The retail price charged by your utility for electricity incorporates not only the fuel used to generate the electricity but the cost of building power plants, administration, lines and substations, and operation and maintenance costs. When you sell your electricity to your utility at retail price, you are being paid more than your electricity is really worth. You could say that your utility is indirectly paying part of the cost of buying PV equipment.

You can find detailed information about net metering in your area by contacting your local utility company or looking at the DSIRE online database (**www.dsireusa.org**). Your utility will ask you to sign a contract, called a Net Metering (NEM) Agreement, spelling out the details of the arrangement. Your electrical contractor will probably also be required to sign. You will be asked to submit the specifications of your PV system and a wiring diagram, and the date when your system will be ready to begin operating. You may be asked to choose between monthly billing or a single annual utility bill. The contract will cover liability and insurance, maintenance, and what happens during power outages. You can view a sample contract for the City of Pasadena at **http://ww2.cityofpasadena.net/waterandpower/solar/PWP_NEM_Final_090208.pdf**.

You will connect to your utility grid through your inverter. All utilities require UL 1741 certification for grid-tie inverters. Each state has utility interconnection standards that dictate the requirements for your grid connection. Your solar contractor or electrician will be familiar with these standards. You will also be required to submit proof that you have all the necessary permits and that your installation conforms to local building codes.

The utility company will inspect your installation to confirm that it conforms with the NEC, the connections are wired correctly, and the workmanship meets certain standards. It will install a new meter and connect it to your inverter.

Renewable Energy Credits and Green Power

The energy you produce with your PV, wind, or biomass system has two kinds of value: the electric power you produce, and the fact that it is clean, renewable energy. Recent climate change legislation mandates that utilities, manufacturing plants, state and local governments, and certain industries reduce their carbon emissions and their consumption of fossil fuels by using renewable energy sources for a certain percentage of their energy needs. In addition, some manufacturers, industries, and organizations want to reduce air pollution and to publicize themselves as "green" industries selling "green" products.

To help in meeting these goals, a system of Renewable Energy Certificates (RECs) has been developed to track the source of energy and to differentiate between renewable energy and energy produced with fossil fuels. One REC represents 1,000 kilowatt-hours of clean, renewable energy. RECs can be bought and sold separately from the electricity they represent and are a sort of energy currency. A utility or industry that does not have access to clean, renewable energy can still meet its goals or declare itself "green" by purchasing RECs from an individual or utility who produces renewable energy. The electricity represented by RECs is known as "green power." RECs are registered, numbered, and tracked by regional tracking systems and any sale of RECs is accompanied by an affidavit that no one else can claim that energy as green.

An REC can be used only once. After a buyer has made an environmental claim based on an REC, that REC is considered permanently "retired." Buyers can also have their RECs retired in their name by their supplier to ensure that no other entity can lay claim to the same environmental benefits.

You own the RECs for the energy produced by your PV system. Some utilities will enter into a separate agreement to buy your RECs, which they can then sell to customers who specifically want "green power." If you sell RECs for electricity generated by your PV system that you use in your own home, you can no longer claim to be using green power.

In Minnesota, Xcel Energy's Solar Rewards program gives customers a one-time payment of $2.25 per installed watt of generating capacity in exchange for ownership of the RECs produced by the systems for 20 years.

Rentals

SolarCity (**www.solarcity.com**), a company operating in Colorado, Texas, Arizona, California, and Oregon, offers SolarLease, a program that lets customers lease solar systems for no money down and a monthly payment. The savings on utility bills are more than enough to cover the monthly payment, and will increase as electricity prices rise.

Lease a solar system

SolarCity (**www.solarcity.com**), a major U.S. solar provider, now offers SolarLease, a program that allows homeowners to lease instead of purchasing a PV system. In place of an initial outlay of $20,000 to $35,000, customers pay a regular monthly fee which is less than the amount of a utility bill. SolarLease contracts also include warranties and service, replacement of defective panels, and free pick up of the solar panels when the lease expires. At the end of the lease, the customer can opt to replace the old panels with the newest equipment, or continue leasing the system in five-year increments. When the home is sold, the owner either transfers the lease to the new owner or prepays it and adds it to the price of the home. SolarCity currently serves Arizona, California, Colorado, Oregon, and Texas with plans to expand into additional states in 2010.

In some states and utility districts, the leasing plan is known as PurePower. A PV unit is installed on a home and the homeowner pays SolarCity the same amount for the electricity generated by the solar panels as for electricity from the utility. When utility prices go up, however, the homeowner continues to pay the lower original rate for solar energy. Utility companies partner with SolarCity to offer these plans because the solar panels relieve some of the demand for electricity from the grid, and customers are satisfied because they know they are using cleaner energy and locking in lower prices.

CHAPTER 9

Starting Up and Maintaining Your System

Photovoltaic equipment has no moving parts and requires very little maintenance. The problems that do occur are usually due to the failure of some electrical part or to the deterioration of connections because of sun, water, wind, or corrosion. Solar panels are expected to remain in service for 25 years or longer, but other components such as controllers and inverters are guaranteed for only ten or 15 years. Small parts like fuses and diodes can be faulty. Snow or dust on your solar array may cause its power output to diminish. Rodents or insects nesting inside electrical container boxes can cause short circuits or chew into cables.

Familiarize yourself with your system's performance. If you know what to expect, regular monitoring will alert you when part of your system begins to malfunction. While your system is new, observe how it functions at different times of the day and under different weather conditions. Modern controllers and inverters have the capacity to store historical performance data that you can analyze with a computer or online. You can also keep a written record in a solar logbook. In the future you will be able to compare the system's performance with past records and detect abnormal declines in its power output. Both controllers and inverters have displays or dials that show how much power is going through the system and alarms that alert you when the system has exceeded one of its limits.

Starting Up Your System

When all the components of your system are fully installed, inspections completed, and (for a grid-tie system) the utility has installed a new meter, you are ready to start generating your own electric power. If your solar system has batteries, it will power up and start charging the batteries as soon as your solar array is connected to your battery controller. A grid-tie system without batteries will start to function when the inverter is turned on.

Read over all the instructions from manufacturers and your utility. You may have to program your solar controller by inputting the battery types and "set points" — minimum and maximum voltage levels that indicate when the batteries are fully charged and when they are depleted. Use the specifications that came with your batteries, or look up their specifications on the Internet.

If your batteries were not fully charged just before you installed them, switch off the inverter and let your system run for 24 hours to charge them up. A new set of flooded (wet) batteries should be fully charged and equalized. Then take a hydrometer reading and record it to use as a baseline for future readings.

Switch your multi-meter to DC voltage and test the output terminals of the battery controller to confirm that it is outputting the correct voltage.

If you have an inverter, turn it on and test it by plugging a small AC appliance such as a lamp into an AC outlet to see if it works. If the appliance does not turn on, switch off the inverter and check the connections to your batteries. Test again with a different small appliance.

Now you are ready to plug in your various appliances and start using clean photovoltaic energy.

System Inspection

Inspect your PV system every six months, in fall and spring.

1. Start with the inverter, if you have one. With a voltmeter and a DC ammeter, verify and record the inverter's operating DC input voltage and current level and its AC output voltage and current levels. Check that the appropriate LEDs are lit up on the inverter to indicate proper operation. If the inverter can show you the total kWh produced since it first started up, make a record of the amount and compare it to your solar array's production since the last inspection. Examine wires and connections for looseness or damage.

2. Examine your solar array for dirt, cracks, signs of leakage, and bent frames. Tighten loose nuts and bolts. Inspect the connections and wiring, check for worn insulation and signs of chewing by rats or squirrels. Replace damaged wiring. Open the junction box(es) and check for dirt and loose connections. Make sure the places where your solar panel mounts penetrate your roof are well sealed and add roof caulk if necessary.

3. Open your connector box and check for dirty, loose, or damaged connections. Measure and record the array's operating voltage and current level on the output side of the combiner box(es) using a voltmeter and DC ammeter. (This should be done in full sun if possible, when you array is producing its maximum output.)

4. Carry out the steps for battery maintenance given below.

Cleaning Your Solar Array

The glass or plastic surfaces of your solar panels should be cleaned at least once a year to remove accumulated dust and grime. Solar industry studies show that panels that do not receive scheduled cleaning lose 15 to 20 percent of their efficiency, increasing payback time by three to five years. Panels that have never been cleaned at all have been found to have lost 30 percent of their efficiency.

Follow manufacturer's cleaning instructions. A simple jet spray with a hose may be sufficient. Do this early in the day before the panels get hot. Rain-repellent glass polish can help to keep dust from accumulating. If you live near a dusty road or construction area, or in a desert area, you will need to wash them more often. In certain seasons, pollen or falling leaves from trees could obstruct your solar panels.

To remove grime and bird droppings, use a solar panel cleaner such as PowerBoost Solar Panel Cleaner (**www.gogreensolar.com/collections/maintenance/products/powerboost-solar-panel-cleaner**) or WinSol Solar Brite (**www.winsol.com/solar.htm**). This can usually be applied with a spray bottle attached to your hose. If you prefer a homemade formula, mix two tablespoons of dishwashing liquid and one tablespoon of vinegar in two cups of water. Use a microfiber cloth or other soft cloth that will not scratch the panels to wipe off excess dust, wipe them gently with the soap mixture, rinse, and dry with another soft cloth. You can buy solar panel cleaning kits with padded, long-handled hose attachments that make it easier to clean hard-to-reach panels.

TIP: Follow safety rules when you are climbing around on a roof.

When cleaning solar panels in high places, always make sure your ladder is anchored firmly at a safe angle. Resist the temptation to over-reach or to balance in an unsafe place to get at difficult corners. Take a few extra minutes to move your ladder and equipment so you can do the job safely.

You can install solar panel self-cleaning devices that regularly spray them with specially formulated soap and filtered water and rinse them. The soap is applied every two weeks and the rinsing is done every two or three days. An example is the Heliotex Automatic Solar Panel Cleaning System (**www.solarpanelcleaningsystems.com/solar-panel-cleaning-services.php**).

If your panels are on an adjustable frame, adjust the tilt seasonally to the optimum angle. Lubricate tracking devices and grease the bearings.

Battery Maintenance

Your batteries require regular attention. Proper maintenance will extend battery life and keep the system performing optimally.

Every three months:

* Observe the display monitor on your controller and see how input from your solar array is matching up with output from your batteries. If the input from your solar array exceeds the output from your batteries, you might have a weak battery or faulty connections between the batteries in your battery pack.
* Check the physical condition of the batteries and battery enclosure. Ventilation areas should be clear, and the enclosure should be dry and insulated from extreme heat or cold.

* Dust off the top of the batteries.

* Look at the battery connectors and make sure they are tightly fitted. For wet batteries, check the fluid levels in the batteries and top up with distilled water if low. (Do this every three months for batteries that are more than five years old.) Use only distilled water. **Never add acid** to a battery except to replace spilled liquid. Water should be added after charging unless the plates are exposed. In that case, add just enough water to cover the plates. After a full charge, the water level should be even in all cells, 1/4" to 1/2" below the bottom of the fill well in the cell (depending on the battery size and type).

* Clean the battery terminals with baking soda dissolved in water, and recoat connections with petroleum jelly.

* If you have multiple wet batteries and your controller has an equalizing function, use it to equalize the batteries. You can equalize the batteries manually by allowing a controlled charge about 10 percent higher than normal full charge voltage for two to 16 hours. This will ensure that all the cells are equally charged, and the gas bubbles will stir up the electrolyte. This is called "boiling" the batteries. AGM and gelled should not be equalized more than two to four times a year. Read manufacturer's recommendations.

Every six months:

* If your controller does not automatically compensate for temperature, and you have wet batteries, adjust the set points for a 12V system to an upper limit of 14.7V for winter and 14.3V for summer. Double these figures for a 24V system, and double them again for a 48V system.

* Use a voltmeter or multi-meter to check the voltage on each battery. They should all be within 0.7 volts of each other. If they are not, follow the steps under "Weak battery" in the section below.

Troubleshooting

A properly designed and installed solar system should give you years of smooth, uninterrupted service. If it does not, you must find and correct the problem. The first signs of trouble are typically intermittent power failures triggered by low battery voltage. The most common problems are:

* You are using more electricity than your system can produce.
* Your system is not producing as much electricity as you expected it to.
* Poor electrical connections or damaged wiring.
* Weak batteries.
* A faulty earth (ground).

Your solar panels, batteries, controller, and inverter all come with manufacturer's instructions, manuals, and often additional support and information on the Internet. Keep this information on file and read over it whenever you have a question or suspect a problem — the solution may be very simple.

TIP: Use safety precautions when troubleshooting.

Observe safety precautions any time you are working with your solar power system. Remember that batteries and solar panels cannot be "switched off." A battery can shock you any time you connect its positive and negative terminals, and a bank of batteries can deliver a powerful current. Solar panels produce power any time they are exposed to light — if necessary cover them with a dark, opaque material while you work on them.

Using too much electricity

The first step in your solar design process was a load analysis of your home to determine how much electricity you use per day. You then created a solar array and battery pack large enough to produce the power you need. You might have overlooked or underestimated some of your loads. The most common reason for solar system failure is that the original design underestimated the amount of power that was needed.

Compare your energy usage with the amount of power your system generates. This information is displayed on the LCD screen of most solar controllers. You can see how much power is being generated, how much energy you are using, and how much charge is in the batteries. Some controllers store and analyze historical data using a computer program or online service, so you can examine your power usage over time. If your system has an inverter, you will also need to study how much power is going into and out of the inverter. Sophisticated inverters, like controllers, provide this information on a screen or through a monitoring program.

If your controller or inverter does not provide power usage information, purchase a multi-meter with the capacity to log data. Attach the multi-meter so it measures the power going from your batteries to your solar controller or inverter and log your power usage over a 24-hour period. Some multi-meters will show you the times of the day when you use the most power.

Solutions

If you determine that you are using more electricity than your system was designed to generate on a daily basis, you have three options:

1. **Reduce or manage your load.**

You might have discovered that a particular appliance such as a heater, air conditioner, washing machine, or refrigerator is placing a heavy load on your system at certain times of the day or year. Depending on the extent to which it drains your system, you can use it less often, replace it with a more energy-efficient alternative, or replace it with an appliance that runs on an independent power source. Remove small energy drains such as digital clocks and cell phone chargers that are plugged in when not in use. Use only CFL bulbs in your lamps.

2. Enlarge your system.

Add more solar panels to your array to generate additional power, and additional batteries to add to your storage capacity. If you were wise and purchased a controller and inverter slightly larger than your system needed, this can be done relatively easily.

3. Add an additional source of power.

Add a generator, fuel cell, or wind turbine to your system to supplement your solar electric system. This may require replacing your controller or inverter.

Your PV array is not producing as much power as it should

Over the years, solar panels begin to degenerate and their productivity gradually decreases. A new system should function smoothly for years before this begins to happen, but obstructions, shade, faulty connections, and failed diodes or fuses can diminish the power produced by your system. Study the input readings on your solar controller several times a day for three to five days to see how much power your solar panels are producing. If your solar controller does not have this function, use a multi-meter to log input from your solar system for several days. If your solar panels are under-performing:

1. Disconnect the solar array from the controller. In full sunlight, connect a voltmeter directly to the solar array or the combiner box and read the open circuit voltage. If it is less than 18V for a 12V array, part or all of your solar array may be defective.

2. Inspect the solar panels for damage, cracked glass or plastic, condensation inside the panels, dirt and grime, bird droppings, and cat paw prints. Wash your panels with warm soapy water and polish with a rain and dirt-repellent glass polish. A damaged panel might have to be replaced, but disconnect it from the system and check its output with a voltmeter before removing it. Even damaged panels often continue to function well, and your real problem might lie somewhere else in the system.

3. Check the wiring for loose connections and corrosion. Repair any connections that are obviously damaged. Replace damaged metal parts. If a terminal on a solar panel is completely destroyed, bypass it by soldering a wire to the metal strip that leads to the solar cells.

4. Do a site survey to see if there are any new obstacles shading your solar array, such as a tree that has grown taller or a neighbor's new chimney. Unfortunately, one of the most common mistakes made by solar installers is failing to note all the obstacles on a site during the design process. Shading may be your problem if the low power output occurs only during certain times of the day. Observe your solar array at those times to see if the sunlight is being obstructed. If shade is your problem, you will either have to remove the obstacle, change the placement of your solar panels to avoid the shade, or add more panels to compensate for the lower power output.

5. Check that your solar array is positioned at the optimum angle to receive the full radiation of the sun at solar noon. If your problem occurs only at certain times of the year, you might need to change the angle to the optimum position for those seasons.

6. If the problem occurs during the summer or at the hottest time of the day, the output of your solar panels may be diminished by heat. Test by cooling your panels with a spray of water. The output should immediately go up. If heat is the problem, you will need to devise a way to cool your solar array.

7. The wiring on some of the cells in a solar panel may be faulty due to poor workmanship or deterioration of the solder. If your array is connected in parallel you can do a selective shading test to find the faulty panel. Enlist an assistant to read the output from your solar array on your controller. In full sunlight, test each solar panel by shading it with an object that covers at least four solar cells. The current coming from your system should drop noticeably when the panel is shaded. If the current does not drop, that solar panel is faulty and is not contributing to the array's power output.

8. PV modules typically have bypass diodes in their junction boxes if they are 24V or higher. A bypass diode protects shaded or weakened cells in a solar panel from overheating by shunting the electric current around them. Occasionally, a diode will short out or fail because of a lightning strike, drastically reducing the output of the PV module. A shorted diode will read near-zero ohms in both directions. If your array is 24V and is not subject to sustained shading, you can simply remove the faulty diode. Otherwise, replace it with a silicon diode with an amps rating at or above the module's maximum current, and with a voltage rating of 400V or more.

Solutions

If your solar panel output is still insufficient after you have taken all the steps suggested above, you have the same three options: reduce your load, enlarge your solar array to compensate, or add another source of power to supplement your system.

Wiring and cables

In full sunlight, with your solar array connected to the system, measure the voltage at the array with a voltmeter, and then measure it at the controller input. A significant drop in voltage means that the resistance in the cable is too high, either because the cable is damaged or because the cable is inadequate for the load it has to carry. Voltage drop will increase as the amount of current flowing through the cable increases. If possible, shorten the length of your cable and test it again. Otherwise, replace it with a larger cable.

- A sudden drop in power during extreme heat or cold could be a sign of damaged wiring or a loose connection either in the solar array or between the solar array and the controller.
- Sudden loss of power when you are running a high load indicates a loose connection in your battery pack or between your batteries and the controller or inverter.
- A significant voltage drop between the output of the controller and your 12V or 24V appliances indicates an inadequate or damaged cable.
- Resistance creates heat. If a cable seems unusually hot to the touch, that cable should probably be replaced immediately. It may be damaged or you may need to use a larger cable.

Check for loose connections, and signs of moisture or corrosion. Confirm that all cables end with a proper terminator or are well soldered. If you suspect poor electrical connections, consult an electrician.

Weak batteries

If your system is not giving you the power you expect, of if your power goes off intermittently when you turn on an appliance, one of your batteries may be faulty. Weak batteries are typically first detected when the weather becomes cold; they can continue operating adequately for months in hot weather.

If you suspect a weak battery, start by performing routine maintenance on your batteries — cleaning, tightening connections, and topping up the fluid levels (in wet batteries). Check the voltage on each of the batteries and use your controller's equalizer or equalize manually if you find a disparity of 0.7 volts or more.

When the plates inside a battery are damaged, sulfated, or partially missing after long use, a battery may give the volt reading of a fully charged battery while behaving like a much smaller battery. Gel batteries can do this if there are bubbles in the gel from overcharging. To get an accurate voltage test, let the battery rest for a few hours first, or put a small load on it such as a lamp.

If one battery's voltage is 2V or more less than the other batteries it is probably faulty. Batteries with failed cells will be unusually hot. Change the faulty battery immediately. Try to find a second-hand battery of the same size, manufacturer, and type, with a similar age and usage level to the other batteries in your pack. If you cannot find a second-hand battery, buy a new battery of the same make and model as the others. You may never get full usage out of this new battery because its performance will match the rest of the batteries. Do not use a new battery in a battery pack that is more than six months old or past 75 cycles; either find a similar used battery or replace the whole battery pack. With some long-life batteries, you can still use a new battery for replacement until the battery pack is one year old. When your batteries are several years old and are near the end of their cycle life, replace the entire battery pack at one time.

When you replace a battery, make sure both it and the other batteries in the pack are all fully charged before reconnecting them. Label the new battery with the date it was replaced.

TIP: Take old lead acid batteries to a scrap yard.

Old batteries can be 100 percent recycled, and a scrap yard may even pay you for them. Do not leave them sitting outdoors or where they can corrode or leak.

Controller

Controllers prolong battery life by regulating the charge coming into the batteries and shutting off outgoing current when the batteries have discharged to a certain point. A malfunctioning controller may damage your batteries. If you are not getting the storage capacity that you expect from your battery bank, or suspect that your controller is malfunctioning:

- Watch your voltmeter as the batteries reach full charge. If the voltage is reaching (but not exceeding) the appropriate set points for your type of battery, the controller is doing its job. If not, read the manufacturer's instructions and try re-programming the set points.
- Look and listen for signs of overcharging, such as severe bubbling in the batteries and moisture accumulating on the battery tops.

Inverter

Many inverters come with diagnostic functions that monitor your PV system and detect problems. Manufacturers also offer diagnosis as part of their warranty and service agreements. Check the manufacturer's manual for troubleshooting suggestions. If you are not getting power from your inverter, there could be several reasons:

* A blown fuse or tripped circuit breaker has cut the current supply to your appliances. Check your fuse box and circuit breakers.
* A ground fault may have caused your inverter to shut off. With the power off, check for and repair ground faults before turning the inverter on again.
* High or low voltage may have triggered one of the inverter's internal disconnects. This will be detected by the inverter's self-diagnosis.
* Your load may exceed your inverter's output. Either reduce the load, or replace the inverter with one that has a higher output.
* The inverter in a grid-tie system senses the utility's voltage and frequency and produces AC electricity at the same voltage and frequency. Low or high utility voltage sensed by the internal disconnects will cause the inverter to shut down. Contact your utility.
* A problem on the battery or array side of the inverter has tripped one of the internal disconnects.

Ground fault

Your PV system has a ground, or earth — a wire that will carry excessive high voltage into the ground in case of a lightning strike or a short circuit.

An incorrectly installed ground or an electrical fault in your system could be causing energy to flow out of the system through the ground instead of into your home.

Conclusion

A smoothly running PV system is hardly noticeable and you can easily forget that it needs regular attention. Keeping your solar panels clean and free of residue will ensure that they maintain their optimum performance and continue to provide you with the power that you expect. Eventually, exposure to sunlight and the elements will cause some deterioration of cables and connections and these need to be repaired or replaced when they are loose or damaged. Your battery system requires supervision to make sure the batteries are fully charging and to extend battery life. You should be able to perform many of these tasks yourself; if not, make arrangements with a solar contractor for regular maintenance. For problems with your inverter or controller, contact the manufacturer or an electrician.

Conclusion

You have learned that solar power, though not a complete solution to the problems created by our use of fossil fuels to produce energy, will play an important role in the future of electrical power. Large-scale solar power plants are already generating electricity for millions of utility customers in several parts of the world, and solar energy is used commercially to power office buildings, warehouses, factories, and agricultural facilities. Though solar energy production currently accounts for less than 0.01 percent of total global primary energy demand, the demand for solar power has grown by 30 percent every year since 1995, compared to less than 2 percent for hydrocarbon energy. Governments are heavily invested in promoting the development and mass-production of new and more efficient solar technologies that will make the use of solar energy more practical and more affordable. Millions of homeowners have already installed residential PV systems that provide all or part of their energy needs. Some of the first systems, which were installed 40 years ago, are still in operation, proving that PV technology lives up to its promises.

The decision you must make is not whether to switch to solar power but when. Prices of components for solar systems have been dropping an average of 4 percent every year for the past 15 years. If you delay your solar project for a year or two, you may have access to new and more efficient technologies that will cut the cost per kWh of your solar electricity. On the other hand, many of the rebates and incentives subsidized by local, state, and federal governments may not be available much longer. Though the economic recession has reduced funding, increased marketing by the solar industry has made these programs so popular that eligible applicants are being turned away. Spain recently made economic news when its government was forced to drastically cut the subsidies that made it the largest market for solar systems during 2008, accounting for half of the world's solar installations. Tax deductions for solar installations will probably remain, but without rebates and other incentives, residential solar installations are still not justifiable for economic reasons alone. If you live in a remote location not yet connected to the power grid, or you are building or remodeling a home, it makes sense to include a PV system in your immediate plans.

To take advantage of rebates and subsidies, and to validate warranties on your solar panels, controllers, and inverters, at least part of your PV system will need to be installed by a licensed contractor or electrician, and inspected for compliance with the NEC. Do not undertake a large do-it-yourself project unless you are building a stand-alone system and do not care about warranties, or you are a licensed electrician. Many smaller installations, such as a system to power a well pump, come as DIY kits. Correct installation is important not only for warranties, but because PV systems carry powerful voltages and can be dangerous for you or your family. Good electrical connections ensure that your system will continue to function smoothly for two or three decades. If you are not sure about what you are doing, take some training classes or consult a professional.

Once your PV system is installed, you will be able to manage and maintain it yourself with relatively little difficulty. You will quickly become an expert at analyzing power usage and battery charge. When attempting repairs and adjustments, always review the manufacturer's instructions so you do not inadvertently void a warranty.

While reading this book, you have probably become aware that you are wasting energy in your home every day, and that there are many small steps you can take to conserve energy and make your home more energy-efficient. This valuable knowledge is essential for heading off potential energy shortages in the future. If every homeowner knew what you know, and consciously reduced his or her consumption wherever possible, millions of kWh would be saved.

You are now one of the millions of scientists, government officials, conservationists, utility managers, and concerned individuals who understand our world energy crisis and the urgency of taking steps to mitigate it. The increasing publicity accorded to clean energy is not just a fad — it is a strategy for human survival. This is only the beginning of a lifelong adventure during which you will see solar technology develop and expand in ways you never dreamed of, and see individuals all over the globe step up and take responsibility for the future of mankind.

APPENDIX

1

Average Monthly Cost to Run Appliances

This chart shows how much it costs to run an appliance per month using electricity supplied by a utility at $0.16 and $0.09 per kWh. Compare your utility's price per kWh.

APPLIANCE	Typical Wattage	Percentage of time appliance is running while turned on	Average Hours Used	Average Monthly kWh	Typical Monthly Cost at $0.16 per kWh	Typical Monthly Cost at $0.09 per kWh
Air Compressor - 1 H.P.	1000	100	20	20	$3.17	$1.80
Air Compressor - 1/2 H.P.	500	100	20	10	$1.59	$0.90
Air Compressor - 3 H.P.	3000	100	20	60	$9.51	$5.40
Air Compressor - 6 H.P.	6000	100	20	120	$19.02	$10.80
Air Conditioner - 10,000 BTU	1000	75	200	150	$23.78	$13.50
Air Conditioner - 5,000 BTU	500	75	200	75	$11.89	$6.75
Air Conditioner - 7,000 BTU	750	75	200	112	$17.76	$10.08
Air Purifier	120	100	720	86	$13.63	$7.74
Answering Machine	10	100	720	7	$1.11	$0.63
Boat De-Icer - 1 H.P.	1000	100	720	720	$114.14	$64.80
Boat De-Icer - 1/2 H.P.	500	100	720	360	$57.07	$32.40

APPLIANCE	Typical Wattage	Percentage of time appliance is running while turned on	Average Hours Used	Average Monthly kWh	Typical Monthly Cost at $0.16 per kWh	Typical Monthly Cost at $0.09 per kWh
Boat De-Icer - 3/4 H.P.	750	100	720	540	$85.61	$48.60
Bug Killer	40	100	300	12	$1.90	$1.08
Ceiling Fan w/o light	50	100	180	9	$1.43	$0.81
Ceiling Fan with 3 - 60 Watt Bulbs	230	100	180	41	$6.50	$3.69
Christmas Lights - 100 Large Bulbs	70	100	150	10	$1.59	$0.90
Christmas Lights - 100 Small Bulbs	50	100	150	8	$1.27	$0.72
Clock - Electric	3	100	720	2	$0.32	$0.18
Clothes Dryer - Electric	5000	100	24	120	$19.02	$10.80
Clothes Washer	1200	100	16	19	$3.01	$1.71
Coffee Maker	900	100	13	12	$1.90	$1.08
Compactor (Trash)	400	100	10	4	$0.63	$0.36
Computer Printer (Printing)	600	100	3	2	$0.32	$0.18
Computer with Monitor	270	100	120	32	$5.07	$2.88
Deep Fat Fryer	1500	100	3	4	$0.63	$0.36
Dehumidifier - 20 pint	480	50	720	173	$27.43	$15.57
Dehumidifier - 40 pint	625	50	720	225	$35.67	$20.25
Dehumidifier - 65 pint	790	50	720	284	$45.02	$25.56
Dishwasher (With Dry Cycle)	1000	50	20	10	$1.59	$0.90
Dishwasher (Without Dry Cycle)	200	100	20	4	$0.63	$0.36
DVD	60	100	120	7	$1.11	$0.63
Electric Blanket	165	50	240	20	$3.17	$1.80
Electric Fence	10	100	720	7	$1.11	$0.63
Electric Frying Pan	1500	100	10	15	$2.38	$1.35
Electrostatic Air Cleaner (On Furnace)	50	100	720	36	$5.71	$3.24
Exercise Equipment - Treadmill 2 H.P.	2000	100	15	30	$4.76	$2.70
Exercise Equipment - Treadmill 3 H.P.	3000	100	15	45	$7.13	$4.05
Fan	200	100	50	10	$1.59	$0.90
Fan (Attic)	500	100	60	30	$4.76	$2.70
Fan (Window)	200	100	50	10	$1.59	$0.90
Fax Machine	10	100	720	7	$1.11	$0.63

Appendix 1: Average Monthly Cost to Run Appliances

APPLIANCE	Typical Wattage	Percentage of time appliance is running while turned on	Average Hours Used	Average Monthly kWh	Typical Monthly Cost at $0.16 per kWh	Typical Monthly Cost at $0.09 per kWh
Fish Tank (10 Gallon)	80	50	720	29	$4.60	$2.61
Fish Tank (50 Gallon)	230	50	720	83	$13.16	$7.47
Freezer - Upright/Chest 17 cu. ft	600	40	720	173	$27.43	$15.57
Freezer - Upright/Chest 17 cu. ft. - Frostfree	600	50	720	216	$34.24	$19.44
GameCube	16	100	120	2	$0.32	$0.18
Grill - Counter Top	1425	100	8	11	$1.74	$0.99
Hair Dryer (Hand Held)	1500	100	10	15	$2.38	$1.35
Heat Lamp	250	100	20	5	$0.79	$0.45
Heater - Auto Engine - 1,000 Watt	1000	50	360	180	$28.54	$16.20
Heater - Auto Engine - 500 Watt	500	50	360	90	$14.27	$8.10
Heater - Portable - 1500 Watt	1500	100	75	112	$17.76	$10.08
Heating Cable - Roof - 60 ft.	500	100	30	15	$2.38	$1.35
Heating Cable - Water Pipes - 24 ft.	200	100	720	144	$22.83	$12.96
Heating System - Hot Air 1/2 HP Motor	500	40	720	144	$22.83	$12.96
Heating System - Hot Air 3/4 HP Motor	750	40	720	216	$34.24	$19.44
Heating System - Hot Water (1 Zone)	315	40	720	91	$14.43	$8.19
Heating System - Hot Water (2 Zones)	423	40	720	122	$19.34	$10.98
Heating System - Hot Water (Summer Use)	135	40	720	39	$6.18	$3.51
Home Theater Receiver	100	100	180	18	$2.85	$1.62
Hot Tub - Insulated/Indoor (4 person)	1500	15	720	162	$25.68	$14.58
Hot Tub - Insulated/Outdoor (4 person)	1500	55	720	594	$94.17	$53.46
Humidifier - Cool Mist	200	100	200	40	$6.34	$3.60
Humidifier - Warm Mist	384	100	200	77	$12.21	$6.93
Iron	1100	50	10	6	$0.95	$0.54
Laptop	75	100	90	7	$1.11	$0.63

APPLIANCE	Typical Wattage	Percentage of time appliance is running while turned on	Average Hours Used	Average Monthly kWh	Typical Monthly Cost at $0.16 per kWh	Typical Monthly Cost at $0.09 per kWh
Lawn Equipment - Hedge Trimmer	450	100	5	2	$0.32	$0.18
Lawn Equipment - Weed Eater	500	100	5	2	$0.32	$0.18
Lawn Mower	3000	100	5	15	$2.38	$1.35
Lighting - 10 rooms (15 60W)	900	100	100	90	$14.27	$8.10
Lighting - 100 Watt	100	100	240	24	$3.80	$2.16
Lighting - 3 rooms (8 60W)	480	100	100	48	$7.61	$4.32
Lighting - 40 Watt	40	100	240	10	$1.59	$0.90
Lighting - 5 rooms (10 60W)	600	100	100	60	$9.51	$5.40
Lighting - 60 Watt	60	100	240	14	$2.22	$1.26
Lighting - 7 rooms (12 60W)	720	100	100	72	$11.41	$6.48
Lighting - 75 Watt	75	100	240	18	$2.85	$1.62
Lighting - Chandelier 5 - 40 Watt Bulbs	200	100	240	48	$7.61	$4.32
Lighting - Comp Fluorescent - 18 Watt	18	100	240	4	$0.63	$0.36
Lighting - Comp Fluorescent - 23 Watt	23	100	240	6	$0.95	$0.54
Lighting - Fluorescent 2 bulb	100	100	240	24	$3.80	$2.16
Lighting - Halogen	90	100	240	22	$3.49	$1.98
Medical Equipment - Nebulizer	1035	100	45	47	$7.45	$4.23
Medical Equipment - Oxygen Concentrator	460	100	720	331	$52.47	$29.79
Microwave Oven	1500	100	10	15	$2.38	$1.35
Mixer - Stand	300	100	20	6	$0.95	$0.54
Motor - 1 H.P.	1000	100	20	20	$3.17	$1.80
Motor - 1/4 H.P.	250	100	20	5	$0.79	$0.45
Motor- 1/2 H.P.	500	100	20	10	$1.59	$0.90
Oven	5000	50	10	25	$3.96	$2.25
PlayStation One	4	100	120	0	$0.00	$0.00
Range - Large Surface Unit	2400	100	10	24	$3.80	$2.16
Range - Small Surface Unit	1200	100	10	12	$1.90	$1.08
Refrigerator - 1.7 cu. ft.	126	33	720	30	$4.76	$2.70
Refrigerator - 14 cu. ft.	226	40	720	65	$10.30	$5.85

Appendix 1: Average Monthly Cost to Run Appliances

APPLIANCE	Typical Wattage	Percentage of time appliance is running while turned on	Average Hours Used	Average Monthly kWh	Typical Monthly Cost at $0.16 per kWh	Typical Monthly Cost at $0.09 per kWh
Refrigerator - 14 cu. ft. - Frostfree	383	33	720	91	$14.43	$8.19
Refrigerator - 17 cu. ft - Frostfree	463	33	720	110	$17.44	$9.90
Refrigerator - 19 cu. ft. - Frostfree	509	33	720	121	$19.18	$10.89
Refrigerator - 21 cu. ft. - Frostfree	572	33	720	136	$21.56	$12.24
Refrigerator - Freezer 21 cu. ft. - Side by Side	783	33	720	186	$29.49	$16.74
Refrigerator - Freezer 24 cu. ft. - Frostfree	653	33	720	155	$24.57	$13.95
Refrigerator - Freezer 25 cu. ft. - Side by Side	841	33	720	200	$31.71	$18.00
Septic Pump	1000	100	40	40	$6.34	$3.60
Slow Cooker	200	100	40	8	$1.27	$0.72
Stereo	75	100	130	10	$1.59	$0.90
Sump Pump	500	100	20	10	$1.59	$0.90
Swimming Pool - Above Ground	500	50	360	90	$14.27	$8.10
Swimming Pool - In Ground 16 × 32	500	50	360	90	$14.27	$8.10
Swimming Pool - In Ground 18 × 36	750	50	360	135	$21.40	$12.15
Swimming Pool - In Ground 20 × 40	1000	50	360	180	$28.54	$16.20
Telephone - Cordless	5	100	720	4	$0.63	$0.36
Television - 13 inch	60	100	120	7	$1.11	$0.63
Television - 19 inch	100	100	120	12	$1.90	$1.08
Television - 25 inch	123	100	120	15	$2.38	$1.35
Television - 27 inch	125	100	120	15	$2.38	$1.35
Television - 32 inch	130	100	120	16	$2.54	$1.44
Television - 36 inch	133	100	120	16	$2.54	$1.44
Television - 43 inch	200	100	100	20	$3.17	$1.80
Television - 55 inch	220	100	120	26	$4.12	$2.34
Television - 60 inch	240	100	120	29	$4.60	$2.61
Television 42" - 50" Projection	320	100	120	38	$6.02	$3.42
Television Cable Converter Box	35	100	720	25	$3.96	$2.25
Television Plasma 42" - 50"	375	100	120	45	$7.13	$4.05
Television/DVD/VCR Combination	120	100	120	14	$2.22	$1.26
Toaster	1000	100	3	3	$0.48	$0.27

APPLIANCE	Typical Wattage	Percentage of time appliance is running while turned on	Average Hours Used	Average Monthly kWh	Typical Monthly Cost at $0.16 per kWh	Typical Monthly Cost at $0.09 per kWh
Toaster Oven	1500	27	25	10	$1.59	$0.90
Tools - Arc Welder	5000	100	5	25	$3.96	$2.25
Tools - Bench Grinder	600	100	10	6	$0.95	$0.54
Tools - Circular Saw	1000	100	10	10	$1.59	$0.90
Tools - Drill	400	100	10	4	$0.63	$0.36
Tools - Saber Saw	400	100	10	4	$0.63	$0.36
Tools - Sander Belt	300	100	10	3	$0.48	$0.27
Tools - Soldering Gun	600	100	10	6	$0.95	$0.54
Tools - Table Saw	3000	100	10	30	$4.76	$2.70
Vacuum - Central	800	100	10	8	$1.27	$0.72
Vacuum - Regular	1440	100	6	9	$1.43	$0.81
Vaporizer	750	100	4	3	$0.48	$0.27
VCR	45	100	30	1	$0.16	$0.09
Video Game	200	100	100	20	$3.17	$1.80
Water Cooler With Hot Water	600	15	720	65	$10.30	$5.85
Water Heating - LCS 10-11 Hours	4500	25	308	346	$54.85	$31.14
Water Heating - LCS 8 Hours	4500	32	243	350	$55.49	$31.50
Water Heating - Master MTRD (20 G)	4500	10	720	324	$51.36	$29.16
Water Heating - Quick Recover (QR)	4500	11	720	356	$48.25	$32.04
Water Pump	900	100	43	39	$6.18	$3.51
Waterbed - Double 100 °	375	74	720	200	$31.71	$18.00
Waterbed - Double 80 °	375	37	720	100	$15.85	$9.00
Waterbed - Double 90 °	375	62	720	167	$26.47	$15.03
Waterbed - King 100 °	375	100	720	270	$42.80	$24.30
Waterbed - King 80 °	375	50	720	135	$21.40	$12.15
Waterbed - King 90 °	375	83	720	224	$35.51	$20.16
Waterbed - Queen 100 °	375	87	720	235	$37.25	$21.15
Waterbed - Queen 80 °	375	43	720	116	$18.39	$10.44
Waterbed - Queen 90 °	375	72	720	194	$30.75	$17.46
Whirlpool Tub	1800	100	15	27	$4.28	$2.43

Acronyms and Abbreviations Used in this Book

A — Amps

AC — Alternating Current

ACAA — American Coal Ash Association

ACH — Air Changes/Hour

AE — Austin Energy

AGM — Absorbed Glass Mat

AH — Amp Hours

AMT — Alternative Minimum Tax

ASES — American Solar Energy Society

ASHRAE — American Society of Heating, Refrigerating, and Air-Conditioning Engineers

ASTM — American Society for Testing and Materials

AWG — American Wire Gauge

BBB — Better Business Bureau

BIPV — Building-integrated Photovoltaics

BPI — Building Performance Institute

Btu — British Thermal Units

C — Centigrade

CC&R — Covenants, Conditions, and Restrictions

CEC — California Energy Commission

CFL — Compact Fluorescent Bulbs

CLFR — Concentrating Linear Fresnel Reflector

CPV — Concentrated Photovoltaic

CSA — Canadian Standards Association

CZTS — Copper, Zinc, Tin, and Sulfur combined with Selenium

DC — Direct Current

DIY — Do-It-Yourself

DOD — Depth of Discharge

DOE — U.S. Department of Energy

DSIRE — Database of State Incentives for Renewables & Efficiency

E — Energy

ECAD — Energy Conservation Audit and Disclosure

EEM — Energy Efficient Mortgages

EERE — Energy Efficiency and Renewable Energy

Eg — Band Gap

EIA — U.S. Energy Information Administration

EPA — Environmental Protection Agency

ETL — ETL Testing Laboratories, Inc.

F — Fahrenheit

FHLMC — Federal Home Loan Mortgage Corporation

GFCI — Ground Fault Circuit Interrupter

GFPD — Ground Fault Protection Device

GLC — General Liability Coverage

GPS — Global Positioning System

GW — Gigawatt

HCPV — High Concentration Photovoltaics

HERS — Home Energy Rating System

HOA — Homeowners' Association

HUD — U.S. Department of Housing and Urban Development

HVAC — Heating, Ventilation, and Air Conditioning

I — Current

IBC — International Building Code

IEC — International Electrotechnical Commission

IECC — International Energy Conservation Code

IEEE — Institute of Electrical and Electronics Engineers

IP — Inverter Priority

IREC — Interstate Renewable Energy Council

IRS — Internal Revenue Service

ISIS — Institute of Science in Society

ITC — Investment Tax Credit

kW — Kilowatts

kWh — Kilowatt hours

LCD — Liquid Crystal Display

LED — Light Emitting Diode

LOI — Letter of Intent

MACRS — Modified Accelerated Cost-Recovery System

MCS — Microgeneration Certification Scheme

MIT — Massachusetts Institute of Technology

MPP — Maximum Peak Power

MPPT — Maximum power point tracking

MW — megawatt

NABCEP — North American Board of Certified Energy Practitioners

NCSU — North Carolina Solar Center at North Carolina State University

NEC — National Electrical Code

NEM — Net Metering

NFPA — National Fire Protection Association

NiCad — Nickel-cadmium

NREL — National Renewable Energy Laboratory

P — Power

PDA — Personal Digital Assistant

PETE — Photon Enhanced Thermionic Emission

PTC — Production Tax Credit

PTC — PVUSA Test Conditions

PURPA — Public Utility Regulatory Policy Act

PVGFPD — PV Ground Fault Protection Device

PVUSA — Photovoltaics for Utility Scale Applications

PWM — Pulse Width Modulation

R — Resistance

REAP — Rural Energy for America Program

REC — Renewable Energy Credit

REPI — Renewable Energy Production Incentive

RESNET — Residential Energy Services Network

SAI — Solar America Initiative

SEIA — Solar Energy Industries Association

SRCC — Solar Rating Certification Corporation

SSL — Solid State Lighting

STC — Standard Test Conditions

TFPV — Thin-film photovoltaic cell

TFSC — Thin-film solar cells

TOF — Tilt and Orientation Factor

TOU — Time of Use

TSRF — Total Solar Resource Fraction

UL — Underwriters Laboratories

UPS — Uninterruptable Power Supply

USDA — U.S. Department of Agriculture

V — Volts

VA — Veterans Administration

VDR — Vapor Diffusion Retarder

Vpp — peak power voltage

Useful Websites

General Information

AltE University. (**www.altestore.com/learn/**)

Average Retail Price of Electricity to Ultimate Customers by End-Use Sector, by State. EIA. (**http://www.eia.gov/electricity/monthly/**)

Carbon footprints of energy production. Parliamentary Office of Science and Technology. (**www.parliament.uk/documents/post/postpn268.pdf**)

Conversion Efficiency. EERE. (**http://energy.gov/eere/energybasics/energy-basics**)

Get Your Power From the Sun (**www.nrel.gov/docs/fy04osti/35297.pdf**)

Homepower Magazine (**http://homepower.com/webextras/**)

How a Passive Solar Home Design Works. DOE. (**http://energy.gov/energysaver/energy-saver**)

How Solar Energy Works. Union of Concerned Scientists. (**www.ucsusa.org/clean_energy/technology_and_impacts/energy_technologies/how-solar-energy-works.html**)

Life-Cycle Environmental Performance of Silicon Solar Panels. (**www.oregon.gov/ODOT/HWY/OIPP/docs/solar_panel_lifecycle.pdf?ga=t**)

Solar Botanic Renewable Energy Systems. (**www.solarbotanic.com/about**)

Solar Energy Global Price Indices. Solarbuzz.com (**www.solarbuzz.com/solarindices.htm**)

Solar Magazine. Solar Server (**www.solarserver.com/solar-magazine/solar-news.html**)

Financing, Rebates, and Tax Incentives

Database of State Incentives for Renewables & Efficiency (DSIRE) (**www.dsireusa.org**)

Does It Pay? Figuring the financial value of a solar or wind energy system. Andy Black. Solar Today. (**www.ases.org/index.php?option=com_content&view=article&id=792&Itemid=23)**

Federal Tax Credits for Consumer Energy Efficiency. Energy Star®. (**www.energystar.gov/index.cfm?c=tax_credits.tx_index**)

IRS Form 5695: Residential Energy Credits (**www.irs.gov/pub/irs-pdf/f5695.pdf**)

P.L. 110-343 | The Emergency Economic Stabilization Act of 2008: Energy Tax Incentives (**http://www.energy.gov/downloads/p**)

Renewable Energy Certificates (**www.epa.gov/greenpower/gpmarket/rec.htm**)

SEIA Tax Manual FAQs (**http://seia.org/galleries/pdf/SEIATaxManual_v3-0_FAQ.pdf**)

SolarLease. SolarCity. (**www.solarcity.com**). PV system leasing.

Take advantage of tax credits and rebates. Consumer Reports. October, 2010. (**www.consumerreports.org/cro/magazine-archive/2010/october/home-garden/saving-energy/energy-tax-credits/index.htm**)

The Borrowers Guide to Financing Solar Energy. NREL. (**www.nrel.gov/docs/fy99osti/26242.pdf**)

Federal, State and Local Agencies, Organizations, Initiatives and Professional Associations

American Solar Energy Society (ASES) (**www.ases.org**)

Building Energy Codes Program. DOE. (**www.energycodes.gov**)

Institute of Electrical and Electronics Engineers (IEEE) (**www.ieee.org**)

IEEE Spectrum: Green Tech (**http://spectrum.ieee.org/green-tech**)

Building Energy Codes Program, DOE (**www.energycodes.gov**)

Interstate Renewable Energy Council (IREC) (**www.eren.doe.gov/irec**)

Million Solar Roofs Final Report 2006 (**www.nrel.gov/docs/fy07osti/40483.pdf**)

National Fire Protection Association (NFPA) (**www.nfpa.org**)

National Center for Photovoltaics (**www.nrel.gov/ncpv**)

North American Board of Certified Energy Practitioners (NABCEP) (**www.nabcep.org**)

North Carolina Solar Center, North Carolina State University (**www.ncsc.ncsu.edu**)

Residential Energy Services Network (RESNET) (**www.resnet.us**)

Solar America Board for Codes and Standards (**www.solarabcs.org**)

Solar Energy Industries Association (SEIA) (**www.seia.org**)

Solar Energy Technologies Program (**www.eere.energy.gov/solar**)

Solar Rating and Certification Corporation (**www.solar-rating.org**) Certification of solar water heaters and swimming pool heaters.

Thin Film Partnership Program. NREL. (**www.nrel.gov/pv/thin_film**)

U.S. Energy Information Administration (**http://www.eia.doe.gov**)

Energy Efficiency

"*An Introduction to Residential Duct Systems.*" The Ernest Orlando Lawrence Berkeley National Laboratory. (**http://ducts.lbl.gov**)

Building Energy Codes Program. DOE (**www.energycodes.gov**)

DOE Zip Code Insulation Calculator (**www.ornl.gov/~roofs/Zip/ZipHome.html**)

Energy Star®. U.S. Environmental Protection Agency and the U.S. Department of Energy. (**www.energystar.gov**)

Calculators and Tools

Cable Resistance Calculator (**www.epanorama.net/index.php?index =calc_cable**)

Compound Angle Tool (**www.solmetric.com/compound-angle.html**)

GetSolar Calculator will show you exactly what cost savings, capital requirements, and investment return you can expect from a solar electric system. (**www.getsolar.com/solar-calculator/index.php**)

Home Power Magazine offers a downloadable Excel spreadsheet with the formulas for calculations already included. (**http://homepower.com/webextras**)

NOAA Solar Calculator (**www.esrl.noaa.gov/gmd/grad/solcalc**) shows paths of the sun at any location in the world.

PVWatts Version 2 Calculator uses hourly typical meteorological year weather data and a PV performance model to estimate annual energy production and cost savings for a crystalline silicon PV system in any location in the United States or its territories. (**www.nrel.gov/rredc/pvwatts/version2.html**)

Single Family Dwelling Electric Load Calculator (**www.electricalknowledge.com/SFDLoadCalc.asp**)

Solar Radiation Monitoring Laboratory at the University of Oregon provides printable sun charts for any location. (**http://solardat.uoregon.edu/SunChartProgram.html**)

Wire Size Calculator (**www.freesunpower.com/wire_calc.php**)

Zip-Code Solar Insolation Calculator (**www.solarpanelsplus.com/solar-calculator**)

Designing Your Solar System

Determining your solar power requirements and planning the number of components. (**www.solar4power.com/solar-power-sizing.html**)

Building Codes Assistance Project. (**http://bcap-energy.org/node/5**)

Energy Matters LLC - Searchable database of 3,720 solar, wind, and renewable energy professionals, accessible at Find-Solar.org (**www.find-solar.org**) and Solar Estimate.org (**www.solar-estimate.org**).

Estimating Appliance and Home Electronic Energy Use. DOE. (**www.energysavers.gov/your_home/appliances/index.cfm/mytopic=10040**)

High-Performance Home Technologies: Solar Thermal & Photovoltaic Systems. Building America Best Practices Series. DOE. (**http://apps1.eere.energy.gov/buildings/publications/pdfs/building_america/41085.pdf**)

How to Use a Multi-Meter Safely. All About Circuits.com. (**www.allaboutcircuits.com/vol_1/chpt_3/9.html**)

Learning About Renewable Energy. NREL. (**www.nrel.gov/learning/ho_photovoltaics.html**)

Power Unit Converter (**www.unitconversion.org/unit_converter/power.html**))

PV Resources.com (**www.pvresources.com/en/pages.php**)

Quick Guide 2 Solar. USDA. (**www.rurdev.usda.gov/or/biz/QuickGuide2SolarPV.pdf**)

Solar Site Survey (**www.builditsolar.com/SiteSurvey/site_survey.htm**)

Batteries

Batteries (**www.solarray.com/TechGuides/Batteries_T.php**)

Battery Tutorial (**www.batterystuff.com/tutorial_battery.html#3**)

Deep Cycle Battery FAQ (**www.windsun.com/Batteries/Battery_FAQ.htm#AGM,%20or%20Absorbed%20Glass%20Mat%20Batteries**)

Controllers

An Introduction to Charge Controllers. Wholesaler Solar. (**www.wholesalesolar.com/Information-SolarFolder/chargecontroller-article.html**)

Increase Solar Charging With An MPPT Power Tracking Charge Controller. Wholesale Solar. (**www.wholesalesolar.com/Windy/MPPT-article.html**)

Inverters

DC Inverters FAQ (**www.solar-electric.com/solar_inverters/inverters_for_solar_electric.htm**)

Learn About Inverters (**http://partsonsale.com/learninverters.htm**)

Solar Panels

Solar Panel Info: What Do All Those Numbers Mean? SolarPowerBeginner.com. (**www.solarpowerbeginner.com/solar-panel-info.html**)

Wires and Cables

Learn About Wiring Solar Panels and Batteries. (**http://partsonsale.com/learnwiring.htm**)

Voltage Drop Chart (**www.solar4power.com/solar-power-volt_drop.html#Volt%20Drop%20Chart**)

Wire Size Calculator (**www.freesunpower.com/wire_calc.php**)

Installation

Build It Solar: The Renewable Energy Site for Do-It-Yourselfers. (**www.builditsolar.com/index.htm**)

Build-Solar-System.org. DIY kits and information. (**www.build-solar-system.org**)

DIY Tutorials (**www.freesunpower.com**)

Find a Solar Installer. NABCEP (**www.nabcep.org/installer-locator**)

How to — Solar Power – Residential and Mobile PV. Alte University (**www.altestore.com/howto/Solar-Power-Residential-Mobile-PV/c30**)

Installing an Off-Grid Inverter. Alte University. (**www.altestore.com/howto/Installing-an-Off-Grid-Inverter/a17**)

National Electrical Code (NEC) (**www.nmsu.edu/~tdi/pdf-resources/pdf%20version%20divided%20PV/NEC/PV-NEC%201.9/DividedNEC-V-1.9pt1.2.pdf**)

Photovoltaic Wiring for Residential Grid-Tied System (**www.walt-n-anne.com/pv/pv_wiring.htm**)

Selecting a Solar Contractor (**www.conservationcenter.org/assets/docs/SelectingASolarContractor_001.pdf**)

Grounding

Grounding vs. Lightning. Trace Engineering Tech Note. (**www.wholesalesolar.com/pdf.folder/Download%20folder/lightning%26grounding.pdf**)

Solar Trackers — Build Your Own

My Homemade Solar Tracker (**http://pages.prodigy.net/rich_demartile**)

DIYDaSolar.com Network (**http://diysolar.dasolar.com**)

LivingonSolar.com (**www.livingonsolar.com/solar-tracking.html**)

Greenwatts (**http://solartracker.greenwatts.info**)

Wind Turbines

American Wind Energy Association (**www.awea.org/faq/rsdntqa.html**)

Wind Maps (**www.windpoweringamerica.gov/wind_maps.asp**)

Wind Turbines (**www.seco.cpa.state.tx.us/re_wind_smallwind.htm**)

Maintenance and Cleaning

Battery maintenance (**www.thesolar.biz/Battery_charging_article.htm**)

Repairing Broken Solar Panels. Otherpower.com (**www.otherpower.com/otherpower_solar_repair.html**)

"Solar Panel Info: What Do All Those Numbers Mean?" Solar Power Beginner.com. (**www.solarpowerbeginner.com/solar-panel-info.html**)

Personal Stories

My First Year with Solar. Steven Johnson. IEEE Spectrum. September, 2010. (**http://spectrum.ieee.org/green-tech/solar/my-first-year-with-solar**)

Some Observations on Photovoltaic Cell Panels. Oliver Seedy (**www.csudh.edu/oliver/smt310-handouts/solarpan/solarpan.htm**)

The Solar Warrior Photovoltaic System (**www.solarwarrior.com**)

Solar Water Heaters and Solar Pool Heaters

Solar Swimming Pool Heaters. DOE. (**www.energysavers.gov/your_home/water_heating/index.cfm/mytopic=13230**)

Solar Vendors

Backwoods Solar: DC Powered Appliances (**www.backwoodssolar.com/catalog/appliances_dc.htm**)

Crown Batteries (**www.crownbattery.com/PDF/Deep%20Cycle%20Batteries.pdf**)

FirstSolar (**www.firstsolar.com/en/index.php**)

InvertersRUs (**www.invertersrus.com/**)

Kyocera (**www.kyocerasolar.com**)

Mitsubishi Thin Film PV Module (**www.mhi.co.jp/en/products/category/solar_power_generating_system.html**)

PowerFilm thin film photovoltaic solar electric modules
(**www.bigfrogmountain.com/PowerFilmSolarProducts.html**)

Realgoods.com (**www.realgoods.com/**)

Solar Direct (**http://solardirect.com/pv/pv.htm**)

Uni-solar (**www.uni-solar.com**)

Watts Up Meters (**www.wattsupmeters.com**)

Glossary

Absorbed Glass Mat (AGM) — Sealed lead acid batteries that do not have a liquid electrolyte.

Acceptance angle — The angle at which the sun's rays naturally fall into a solar panel or solar concentrator.

Air-based solar heating — A heating system that heats air in an air collector.

Alternating current (AC) — An electric current that reverses itself at regular intervals.

American Solar Energy Society (ASES) — A national nonprofit association of solar professionals and advocates.

Ampere — The unit of measurement used for electric current.

Array — A panel formed by wiring together a number of PV modules.

Arsine — A colorless, flammable, highly toxic gas generated when metals or ores containing arsenic are treated with acid.

Avoided cost — A utility's wholesale cost to produce electricity, typically about ¼ of the retail price.

Azimuth — A horizontal coordinate for locating the sun in the sky, expressed as the number of degrees clockwise from true north.

Bandgap (Eg) — The amount of energy required to release an electron from the silicon in a PV cell.

Baseline usage — A household's rate of electricity consumption with heating and air conditioning factored out.

Batch collector — Solar water heater that heats water in dark tubes in an insulated box.

Blemished solar panel — A new solar panel with a slight defect that does not affect its performance.

Blower door — A powerful fan that lowers air pressure inside of a building to aid in detecting air leaks.

British thermal unit (Btu) — A measure of energy equal to 250 calories, the amount of energy required to raise the temperature of a pound of water by 1° Fahrenheit.

Building-integrated photovoltaics (BIPV) — The integration of PV cells in architectural features and building materials.

Canadian Standards Association (CSA) — The Canadian body that certifies solar panels and equipment.

Carbon footprint — A measure of the impact a person makes on the environment, or the total amount of CO_2 and other greenhouse gases emitted over the full life cycle of a process or product, expressed as grams of CO_2.

Carbon payback — A measure of the way in which a solar system compensates for its carbon footprint.

Cascade cells — See multijunction PV cells.

CEC — California Energy Commission.

Charge Cycle Efficiency — A measure of the percentage of energy you can draw from the battery compared to the amount of energy used to charge it.

Clean energy — A relative term for energy produced without creating pollution, waste, or environmental toxins.

Commutator — A device that regularly reverses the direction of electric current in an electric motor running on DC.

Concentrated photovoltaic arrays — Devices which use lenses or mirrors to focus sunlight onto the solar cells.

Concentrating PV (CPV) — Solar power systems that use optical devices to focus a large amount of solar energy on a small area.

Concentrating solar power (CSP) — Technology in which mirrors or lenses focus and amplify the sun's energy before it is converted to electricity through the photovoltaic process or a thermodynamic heat cycle using a motor.

Conduction — The movement of heat or energy through a solid material.

Convection — The circulation of heat through liquids and gases.

Conversion efficiency — The percentage of the solar energy shining on a PV device that is converted into electrical energy, or electricity by a PV cell.

Covenants, Conditions, and Restrictions (CC&Rs) — The rules and regulations governing members of a homeowners' association.

CIGS — A semiconductor made of copper, indium, gallium, and selenium.

CZTS — A semiconductor material made of copper, zinc, tin, and sulfur and selenium.

Depth of discharge (DOD) — The degree to which a battery is discharged during each cycle.

Direct current (DC) — Electric current that always flows in the same direction.

Dish Stirling — A PV system that focuses solar energy to generate heat for a steam engine that produces electricity.

DOE — U.S. Department of Energy.

Dopant atoms — Atoms with odd numbers of electrons in their outer orbitals that are introduced into semiconductors to initiate a flow of electrons when energy is applied.

Dry cell battery — Another name for AGM or gel batteries.

Duct sealant (duct mastic) — A material used to seal leaks in air conditioning ducts.

Duty cycle — The period during which an appliance controlled by a thermostat or timer is running.

Elevation — A vertical measurement of the sun's position in the sky, measured in degrees as the angle between the horizon and a line drawn through the sun.

Evacuated tube solar collector — Solar water heater that uses transparent glass tubes containing pipes covered with absorbent materials.

Evapotranspiration — The process by which a plant actively moves and releases water vapor.

Federal Public Utility Regulatory Policy Act of 1978 (PURPA) — An act allowing businesses and individuals to sell excess electricity generated by solar or wind systems to their local utilities.

Flat-plate collector — A solar water heater consisting of copper tubes fitted to absorber plates.

Forced hot air heating system — A conventional heating system that uses a furnace to heat the air and circulates it through the home via ducts.

Ground Fault Circuit Interrupter (GFCI) — A device that monitors the current in a circuit entering and leaving an electrical device and shuts off power when a short circuit is detected.

Gigawatt — 1,000 Megawatts.

General liability coverage (GLC) — Insurance that protects you from damage claims due to any eventuality not covered by your homeowner's insurance.

Green power — Electricity certified as being produced by clean, renewable sources.

Grid — A network of electricity providers and consumers connected by transmission and distribution lines.

Grid fallback system — A PV system in which power generated by a solar array is stored in batteries that supply electricity to a home, and power is purchased from a utility only when the batteries are depleted.

Grid failover system — A PV system that stores solar power in batteries for use only when the grid experiences a power outage.

Grid-tie system — A PV system connected to a power grid.

HVAC — Heating, Ventilation, and Air Conditioning

High concentration photovoltaics (HCPV) — PV systems employing solar concentrators that concentrate sunlight to intensities of 300 suns or more.

Hybrid solar energy system — A system in which the natural collection of the sun's energy is enhanced by the use of mechanical devices such as fans or pumps.

Incandescent bulb — A light bulb that contains a tungsten filament.

Insolation — The number of hours of sunlight combined with the strength of the sunlight falling on one square meter (9.9 square feet).

Integrated collector storage systems — Another name for batch collector solar water heaters.

International Energy Conservation Code (IECC) — The most commonly adopted model energy code for residential buildings, updated every three years.

Inverted cell — A solar cell grown on a gallium arsenide wafer, then flipped over and removed from the wafer.

Inverter efficiency — The percentage of DC that an inverter puts out as AC under various conditions.

Life in Float Service — The maximum shelf life of a battery.

Light emitting diode (LED) — A device that emits photons of light when it receives electric energy.

Kilowatt peak — The maximum energy produced by a PV generator at maximum solar radiation under Standard Test Conditions (STC).

Liquid-based solar system — A system that heats either water or an anti-freeze solution in a liquid-based collector.

Load — Anything that draws power from an electrical circuit.

Manufacturer's Certification Statement — A signed statement from the manufacturer certifying that a product or component qualifies for the Residential Renewable Energy Tax Credit.

Megawatt — 1,000,000 Watts

Million Solar Roofs Initiative (MSR) — An initiative launched in June, 1997, by the U.S. Department of Energy (DOE) to have one million solar roofs in place in the United States by 2010.

Microgeneration Certification Scheme (MCS) — A certification program for solar equipment in the United Kingdom.

Modified sine wave — The modified AC sine wave put out by less-expensive inverters.

Multijunction PV cell — A stack of two or more PV cells with different bandgaps and more than one junction.

Multimeter — A meter that measures voltage, current, resistance, and often other electrical variables.

Multitester — Another name for a multimeter.

Nanowire — A wire with a diameter of 10^{-9} meters.

NASA — National Aeronautics and Space Administration

National Electrical Code (NEL) — The code that sets standards for safe electrical installations in the United States.

National Fire Protection Association (NFPA) — A national organization that provides electrical and building codes and standards, research, training, and education to reduce the danger of fire and other hazards.

Net metering — An arrangement through which a local utility buys excess power generated by a solar system, and sells it back when the solar system is inactive.

NiCad batteries — Nickel-cadmium batteries.

NOAA — National Oceanic and Atmospheric Administration

Nonrenewable energy — Energy from sources that cannot be replaced when they are used up.

NREL — Department of Energy's National Renewable Energy Laboratory.

Off-grid system — A stand-alone PV system that is not connected to a power grid.

Ohm — A measure of electrical resistance.

Ohmmeter — An instrument for measuring electrical resistance.

Ongrid system — A PV system connected to a power grid.

Organic PV cells — Solar cells consisting of carbon-containing molecules, known as polymers, that can be sprayed or printed onto a flexible surface.

Peak power rating — The maximum amount of power that an inverter can supply in short bursts.

PETE (photon enhanced thermionic emission) — A process in which both light and heat produce electric current in a photovoltaic process.

Phosphine — A flammable, toxic gas released when phosphorous is used in the manufacture of solar panels.

Photovoltaic cell — A small unit containing silicon that converts sunlight into electricity.

Photovoltaic process — The process by which photons of sunlight release electrons in a photovoltaic cell to produce an electric current.

Photovoltaic module — A unit of photovoltaic cells that are electrically connected and placed in a frame.

Polymer — A large molecule composed of repeating structural units.

Power rating — The maximum continuous power that an inverter can supply to all the loads in a home.

Power tower — A tower with a receiver on top that captures solar energy focused on it by mirrors surrounding the tower, typically as heat that is used in a steam turbine to produce electricity.

PURPA — Federal Public Utility Regulatory Policy Act of 1978.

Quad — One quadrillion Btus.

Quasi-sine wave — The modified AC sine wave signal put out by less-expensive inverters.

Renewable energy — Energy from sources that are constantly replenished, such as sunlight and sustainable forests.

Renewable Energy Certificate (REC) — A certificate representing 1,000 kilowatt-hours of clean, renewable energy.

Resistance — The degree to which a substance opposes the passage of electric current through it.

Sealed regulated valve battery — Another name for an AGM battery.

Semiconductor — A substance that conducts electricity only under certain circumstances, such as when it absorbs radiation from infrared (IR), visible light, ultraviolet (UV), or X rays.

Set points — Minimum and maximum voltages programmed into a battery controller to let it know when the batteries are fully charged or discharged.

Solar America Initiative (SAI) — An initiative launched in 2006 by the U.S. Department of Energy (DOE) which aims to achieve cost parity with conventional electricity generation by 2015.

Solar thermal energy — The use of heat from the sun to provide hot water or a heating system for a building.

Standard Test Condition (STC) — The standard conditions for testing PV products, defined by IEC 60904–3 as an insolation of 1000W/m2 (1 SUN) at 25°C and with a solar spectral distribution equivalent to global AM1.5.

Starved electrolyte batteries — AGM batteries.

String — Two or more solar panels wired together in series.

Sun — A unit of measurement for concentrated solar radiation.

Tandem cells — Another name for multijunction PV cells.

Temperature coefficient of power — The percentage of total power reduction for each 1°C increase in the temperature of a solar panel.

Tempering valve — A valve used with a batch collector solar water heater to reduce the temperature of hot water before it reaches a faucet.

Therm — A unit of heat equal to 100,000 British thermal units.

Thermal capacitance — The ability of materials to store heat.

Thermal mass — Building materials that absorb heat.

Thermodynamic heat cycle — A repetitive energy cycle in which heat is converted to electricity or mechanical energy.

Thermography — The mapping of temperature differences in a substance or on a surface, such as the outer walls of a house during winter.

Thin-film photovoltaic cell (TFPV) — Another name for a thin-film solar cell.

Thin-film solar cell (TFSC) — A solar cell made by depositing one or more thin layers of photovoltaic material on a metallic substrate.

Time of Use (TOU) — A utility price structure that charges more for electricity used during peak hours.

Total solar resource fraction (TSRF) — A measure of the effect of shading on a solar panel system's potential to generate electricity.

Traction batteries — Another name for wet batteries.

Translucent photovoltaics — Another name for transparent photovoltaics.

Transparent photovoltaics — A type of solar cell incorporated in a window or skylight that allows half of the sunlight that falls on it to pass through, and produces electricity using ultraviolet radiation in addition to visible and infrared light.

Underwriters Laboratories (UL) — A national testing facility that certifies solar panels and equipment.

Uninterruptable Power Supply (UPS) — A battery back-up that provides continuous electrical power for computers or machinery during a power failure.

U.S. Department of Energy (DOE) — A Cabinet-level department of the U.S. government that formulates and implements policies regarding energy and the safe use of nuclear power.

UTC (Coordinated Universal Time) — Formerly Greenwich mean time, the global standard for time and date.

Volt — The Standard International (SI) unit of electric potential or electromotive force. A potential of one volt appears across a resistance of one ohm when a current of one ampere flows through that resistance.

Voltage — A measure of the amount of energy contained in an electric field at a given point.

Volt/ohm meter (VOM) — A meter that measures multiple electrical variables, including voltage, current, and resistance.

Watt — A unit of power equal to 1 joule per second.

Wet batteries — Batteries that contain a liquid electrolyte.

Wind chill — The lowering of cold outside temperatures by wind speed.

Bibliography

"*Acute power shortage throws life out of gear.*" The Times of India. May 14, 2010. (**http://timesofindia.indiatimes.com/city/varanasi/Acute-power-shortage-throws-life-out-of-gear/articleshow/5931855.cms**)

Anderson, Jackie and Allard Beutel. "*Solar Power Plant at Kennedy Supplying Electricity to Floridians.*" Kennedy News RELEASE : 20-10. April 8, 2010. (**www.nasa.gov/centers/kennedy/news/releases/2010/release-20100408c.html**)

Austin Energy Solar PV Rebate Program Procedures (**www.austinenergy.com/Energy%20Efficiency/Programs/Rebates/solar%20rebates/solarRebateProcedures.pdf**)

Beaty, William J. "What is Electricity." 1996. (**http://amasci.com/miscon/whatis.html**)

Boxwell, M. *Solar electricity handbook: a simple, practical guide to solar energy — designing and installing photovoltaic solar electric systems.* Ryton on Dunsmore, Warwickshire, U.K., Greenstream Publishing. 2010.

Brain, Marshall. "How Electric Motors Work." How Stuff Works.com. (**http://electronics.howstuffworks.com/motor2.htm**)

"DC Inverter FAQ." Arizona Wind and Sun, Inc.. February 5, 2009. (**www.solar-electric.com/solar_inverters/inverters_for_solar_electric.htm**)

"Electric Motors. Explain That Stuff!" November 26, 2009. (**www.explainthatstuff.com/electricmotors.html**)

"Electric Motors." Solar Navigator. Max Energy Limited. (**www.solarnavigator.net/electric_motors.htm**)

Gulliksen, Josie. "Solar Power in U.S. Homes Saves Money in the Short Term, Too." Housing Watch. July 6, 2010. (**www.housingwatch.com/2010/07/06/solar-energys-becoming-more-prevalent-in-u-s-households**)

Harris, Tom. "How Light Bulbs Work." HowStuffWorks, a Discovery Company. (**http://home.howstuffworks.com/light-bulb2.htm**)

"*Home solar panels doubled electric output last year.*" USA Today GreenHouse. April 19, 2010. (**http://content.**

usatoday.com/communities/greenhouse/post/2010/04/homes-solar-panels-doubled-electric-output-last-year/1)

"*How Much Do Solar Panels Cost?*" Cost Helper.com. February 2007. (**www.costhelper.com/cost/home-garden/solar-panels.html**).

"Increase Solar Charging With An MPPT Power Tracking Charge Controller." Wholesale Solar. (**www.wholesalesolar.com/Windy/MPPT-article.html**)

Lacey, Stephen. "Is Organic PV the Future of Solar?" RenewableEnergyWorld.com. May 3, 2010. (**www.renewableenergyworld.com/rea/news/podcast/2010/05/is-organic-pv-the-future-of-solar-maybe-**)

"*New Solar Energy Conversion Process Could Double Solar Efficiency of Solar Cells.*" ScienceDaily. August 2, 2010. (**www.sciencedaily.com/releases/2010/08/100802101813.htm**)

"*Outrage over power shortage in freezing weather.*" Dawn.com. January 17, 2010. (**www.dawn.com/wps/wcm/connect/dawn-content-library/dawn/news/pakistan/metropolitan/03-pakistanis-furious-as-winter-power-shortages-grip-nation-ss-01**)

Pinkham, Linda. "What's the Average Cost to Install a Solar-Electric System to Power Your Home?" Mother Earth News. August 5, 2010. (**www.motherearthnews.com/ask-our-experts/solar-electric-system-cost-z10b0blon.aspx**)

Solar Electric Tech Tips. Northern Arizona Wind and Sun. (**www.windsun.com/General/tech_tips.htm**)

"Solar Panel Info: What Do All Those Numbers Mean?" Solar Power Beginner.com. (**www.solarpowerbeginner.com/solar-panel-info.html**)

"Solar Retrofits for Weatherization and Remodels." Learning Tool for Low-Income Renovation Providers in Pennsylvania. Pennsylvania Weatherization Providers Task Force. Lesson 8: PV System Maintenance. (**www.pasolar.ncat.org/lesson08.php#inside**)

The Borrowers Guide to Financing Solar Energy. National Renewable Energy Laboratories. (**www.nrel.gov/docs/fy99osti/26242.pdf**)

Toothman, Jessika and Scott Aldous. "How Solar Cells Work." How StuffWorks.com. (**http://science.howstuffworks.com/environmental/energy/solar-cell2.htm**)

U.S. Department of Energy, Energy Efficiency and Renewable Energy. *The History of Solar* (**www1.eere.energy.gov/solar/pdfs/solar_timeline.pdf**)

West, Larry. "Compact Fluorescent Light Bulbs: Change a Light Bulb and Change the World." About.com. (**http://environment.about.com/od/greenlivingdesign/a/light_bulbs.htm**)

Index

G

H

I

K

L

M

N

P

R

S

T

U

V

W

Y